CAMBRIDGE PHYSICAL SERIES.

ELECTRICITY

AND

MAGNETISM

ELECTRICITY

AND

MAGNETISM

AN ELEMENTARY TEXT-BOOK
THEORETICAL AND PRACTICAL

BY

R. T. GLAZEBROOK, M.A., F.R.S.

DIRECTOR OF THE NATIONAL PHYSICAL LABORATORY,
FELLOW OF TRINITY COLLEGE, CAMBRIDGE.

CAMBRIDGE:
AT THE UNIVERSITY PRESS
1903

CAMBRIDGE
UNIVERSITY PRESS

University Printing House, Cambridge CB2 8BS, United Kingdom

Cambridge University Press is part of the University of Cambridge.

It furthers the University's mission by disseminating knowledge in the pursuit of
education, learning and research at the highest international levels of excellence.

www.cambridge.org
Information on this title: www.cambridge.org/9781316626153

© Cambridge University Press 1903

First published 1903
First paperback edition 2016

A catalogue record for this publication is available from the British Library

ISBN 978-1-316-62615-3 Paperback

PREFACE.

SOME words are perhaps necessary to explain the publication of another book dealing with Elementary Electricity. A considerable portion of the present work has been in type for a long time; it was used originally as a part of the practical work in Physics for Medical Students at the Cavendish Laboratory in connexion with my lectures, and was expanded by Mr Wilberforce and Mr Fitzpatrick in one of their Laboratory Note-books of Practical Physics.

When I ceased to deliver the first year course I was asked to print my lectures for the use, primarily, of the Students attending the practical classes; the lectures on Mechanics, Heat and Light have been in type for some years. Other claims on my time have prevented the issue of the present volume until now, when it appears in response to the promise made several years ago.

Meanwhile the subject has changed; but while this is the case the elementary laws and measurements on which the science is based remain unaltered, and I trust the book may be found of service to others besides my successors at the Cavendish Laboratory.

As in the other books of the Series, I have again to thank Mr Fitzpatrick for his very valuable assistance.

He has read all the proofs and suggested numerous improvements, and has thus brought the book up to date as representing a course which many years' experience has proved to be a useful one for elementary students.

The book is to be used in the same way as its predecessors. The apparatus for most of the Experiments is of a simple character and can be supplied at no great expense in considerable quantities.

Thus the Experiments should all, as far as possible, be carried out by the members of the class, the teacher should base his reasoning on the results actually obtained by his pupils. Ten or twelve years ago this method was far from common; the importance to a School of a Physical Laboratory is now more generally recognized; it is with the hope that the book may be of value to those who are endeavouring to put the method in practice that it is issued now.

R. T. GLAZEBROOK.

National Physical Laboratory.
    July 19, 1903.

CONTENTS.

ELECTRICITY.

CHAPTER I.

1. Electric Attraction. The word electricity is derived from ἤλεκτρον the Greek for amber; if we take a piece of sealing-wax or a glass rod and rub it with dry flannel or silk it will be found to have acquired the power of attracting light objects such as bits of tissue paper or feathers. This property of attraction was first discovered by the Greeks in amber; hence when Dr Gilbert about 1600 found that numerous other substances possessed it when rubbed he called them all "Electrics."

This attraction may be illustrated in various ways. Take a dry glass rod and rub it with a dry silk handkerchief, then hold it over a number of bits of tissue paper; the tissue paper jumps up to the rod; probably some bits stick to the rod while others after touching it are repelled and fly away. The same effect can be shewn by rubbing a piece of sealing-wax or a rod of ebonite with dry flannel or catskin.

We can however study the effects better thus. Suspend a light pith ball or a small bit of feather by a thin silk thread; then rub a glass rod with silk and bring it

Fig. 1.

near the ball; the ball is attracted to the glass as in
Figure 1; allow the two to come into contact, the ball is
then repelled from the glass.

Again take a second pith ball suspended in the same way
and rub a rod of ebonite with dry flannel; then bring
the rod near the ball, it is first attracted and then after
contact is repelled from the rod.

The glass and the ebonite have both been electrified by
friction, the first ball has been in contact with the glass, the
second with the ebonite; the glass and the ebonite are both
said to be electrified. Each has the power of attracting to
itself a light body such as the pith ball and then repelling
it. A body which has been electrified is said to be charged
with electricity or to have an electric charge.

2. Two kinds of electrification. Many other
substances can be electrified by friction, such as fur, wool,
ivory, sulphur and india-rubber, but the states of electrification
produced in these various bodies are not the same; we can
shew in fact that there are two kinds of electrification.

EXPERIMENT 1. *To shew that there are two kinds of
electrification produced by friction.*

A wire stirrup is suspended by means of a fine silk fibre
as shewn in Figure 2. The stirrup is intended to support

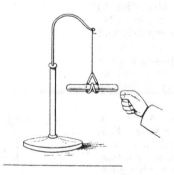

Fig. 2.

a rod of glass or of ebonite after it has been electrified. Rub one end of an ebonite rod with a piece of dry flannel and support it in the stirrup so that it hangs horizontally. Electrify in the same manner one end of a second ebonite rod and bring it near the electrified end of the suspended rod ; the latter is repelled. Now electrify a glass rod by rubbing it with silk and bring it near the electrified end of the suspended ebonite rod ; the ebonite rod is attracted. Thus while two ebonite rods when electrified repel each other, an ebonite rod and a glass one attract ; the effects on the suspended ebonite rod of the glass and ebonite are opposite ; we may express this by saying that the two states of electrification are of opposite sign ; it is usual to call the electricity produced on glass by friction with silk positive, that produced on ebonite or sealing-wax when rubbed with wool or with catskin is called negative; these signs are matters of convention, the electricity on the glass might have been called negative, that on the ebonite positive, had this convention been agreed to instead of that which has actually been adopted.

Thus we have seen from the above experiment that there are two opposite states of electrification, and further that two bodies similarly electrified repel each other, while two bodies oppositely electrified attract each other.

We could vary the above experiment by electrifying a piece of glass by silk and supporting it in place of the ebonite ; we should then find that it was attracted by electrified ebonite, repelled by electrified glass.

Moreover if the silk and the catskin be very dry we can shew in the same way that they also are electrified by friction, for on holding the catskin near to the suspended glass it is attracted while the silk which was used to rub the glass is repelled. Thus the catskin is positively electrified by rubbing ebonite, and the silk negatively by rubbing glass.

3. Conductors and non-conductors. If an electrified piece of ebonite be laid on another piece of ebonite or on a sheet of dry glass or a piece of dry silk, it will retain its electrification for some time ; if however it be rubbed gently with a damp cotton cloth or with the hand or touched

all over with a piece of tin-foil or other metal, it at once
loses its electrification. The ebonite, the glass, and the silk
are said to be non-conductors of electricity; the damp cloth,
the hand, and the metal are called conductors of electricity.
A conductor allows the free passage of electricity over its
surface; if we hold a piece of brass tube in the hand
and rub it with catskin or flannel, no sign of electrification
is produced; if however we fix the brass tube on to an
ebonite rod and rub it with dry catskin taking care to avoid
touching the brass with anything but the fur, we find that
the brass is negatively electrified; if brought near an ebonite
rod suspended as in Experiment 1 it will repel it. The brass
tube, the human body, and the earth are all conductors;
some electricity is produced by the friction in the first case,
but it spreads itself over the brass and passes through the
body of the experimenter to the earth; in the second case
also electricity is produced in the brass but is prevented
from passing to the earth by the ebonite handle. Thus the
tube remains electrified.

There is this difference however between it and the ebonite;
that part only of the ebonite which has been rubbed is electrified, while the whole of the brass tube shews signs of
electrification. Electrification produced by friction on a non-
conductor remains where it was produced, while on a conductor
it spreads itself over the whole surface.

This can be shewn in the following manner:—

On dusting a body with a mixture of powdered red lead
and yellow sulphur which has been well shaken together, both
powders become electrified by the friction, the lead being
positive and the sulphur negative; if the body be unelectrified
both powders can be easily removed by blowing on the surface;
if the body be positively electrified the sulphur is held by it
and the lead is removed by blowing, the surface becomes yellow
when blown upon. If on the other hand the body be
negatively electrified, the red lead is retained and the sulphur
blown off, the surface becomes red; we have thus an easy
test of the nature of the electrification of a body.

Now support a rod of ebonite and a brass tube, each in
some non-conducting support. Rub one end of the ebonite

with catskin and dust it with the red lead-sulphur mixture; the end which was rubbed becomes red, neither powder adheres to the other end, the ebonite is only electrified where it was rubbed. Repeat the experiment with the brass tube, it becomes red all over, the electrification has distributed itself all over the brass.

DEFINITION. *A substance which is a non-conductor of electricity is said to be an* **Insulator** *or to insulate, and a body supported by an insulator is said to be insulated.*

4. Properties of Insulators. Bodies can be divided roughly into two classes, conductors, which allow the free transference of electricity from point to point, and non-conductors or insulators in which that transference takes place very slowly indeed if at all. A perfect insulator would be a body which entirely stopped the passage of electricity; substances ordinarily classed as insulators, glass, shellac, silk, etc., are not perfect, but while in conductors the transference is so rapid as to be practically instantaneous, in a good insulator it is very slow. Indeed glass when thoroughly dry, fused quartz, paraffin wax and various other substances are practically perfect insulators. Water as we know it is a good conductor, though it appears probable that perfectly pure water would be an insulator. Air and the other permanent gases when dry are insulators, cotton thread especially if slightly damp is a conductor, silk is an insulator.

5. Electrification by induction. Bring an electrified ebonite rod near to a conductor supported on an insulating stand, and dust the conductor with the red lead and sulphur powder as described in Section 3; on blowing on the conductor the end near the ebonite will become yellow, the other end red; the nearer end has become positively electrified, the further end negatively. On removing the ebonite the electrification disappears. The conductor is said to be electrified by induction. If a positive body had been used instead of the ebonite to produce the induction, the nearer end would have been negative, the further end positive.

In the above experiment the state of induction lasts only so long as the charged body is near the conductor. We can however produce by induction a permanent state of electrification thus.

EXPERIMENT 2. *To electrify a conductor by induction.*

Bring an electrified ebonite rod near to an insulated conductor, and while the ebonite is in position touch the conductor for a moment with the finger. Then remove the ebonite; on dusting the insulated conductor with the lead-sulphur powder and blowing on it, it will become yellow shewing that it is positively electrified; if a positively charged body had been used instead of the ebonite to produce the induction, the insulated conductor would at the end of the experiment remain negatively electrified, it would thus appear red. Moreover in both these cases there is attraction between the charged rod and the insulated conductor.

To shew this, charge one end of the ebonite rod and suspend it in the stirrup as in Experiment 1. Then bring up near to it the insulated conductor on its stand, the charged end of the rod is attracted. If the conductor is not insulated the attraction is more marked; this can be shewn by touching the insulated conductor with the hand when near the rod, the rod is drawn still closer to the conductor.

We have already seen that on the insulated conductor there is positive electrification near the ebonite, and negative at a distance from it; this enables us to explain the attraction, for we may imagine the conductor divided into two parts, one of these is positively electrified, the other negatively. There is attraction between the ebonite and the positive part, repulsion between the ebonite and the negative part. But the distance between the first two being less than that between the second two, the attraction is greater than the repulsion and hence on the whole the ebonite is attracted to the conductor.

6. Explanation of electrical attraction. We may use these results to explain the phenomena of attraction and repulsion described in the first section, thus:

The bits of paper, the feather, and the pith balls are all of them conductors. When the charged ebonite rod is brought near the pith ball electric induction takes place, the ball becoming positively electrified on its side nearest the ebonite and negatively electrified on the opposite side; the ball is thus attracted by the ebonite; on contact the positive electrification of the ball is discharged by combination with some of the negative electrification of the ebonite, the induced negative charge remains on the ball; hence the ball and the ebonite being similarly charged repulsion takes place and the ball flies away from the ebonite.

7. Electroscopes. An instrument for detecting whether a body is electrified or not is called an electroscope.

In an electroscope as a rule the attraction or repulsion between two electrified bodies is made use of to determine the presence of electrification. Thus the pith ball suspended by a silk fibre might be used as an electroscope; a more sensitive electroscope however can be constructed as is shewn in Figure 3.

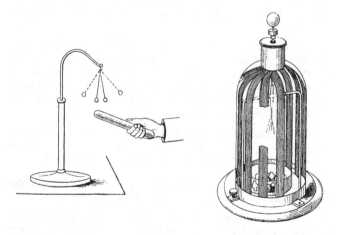

Fig. 3.   Fig. 4.

Two pith balls are connected by a linen or cotton thread and suspended from a glass support. Cotton or linen is used because it is a conductor so that any charge communicated to either ball is shared by the other. If the two balls become electrified they repel each other and stand apart. A number of observations might be made with such an electroscope. The gold-leaf electroscope however, Figure 4, which acts on the same principle, is much more delicate, the balls are replaced by two strips of gold or aluminium leaf. These hang from a metal rod inside a glass vessel. The rod is inserted into a glass tube which is well coated with shellac varnish and passes through a cork or stopper in the neck of the vessel. The shellac improves the insulating properties of the glass greatly[1]. The metal rod terminates in a metal plate or knob which is thus insulated but is in connexion with the gold leaves. For accurate work the surface of the glass vessel is partly covered with strips of tin-foil, spaces being left between the strips through which the gold leaves can be seen, or in some cases a cylinder or cage of wire gauze is placed inside the vessel. The object of these precautions is to prevent electrification on the surface of the glass or on external objects from influencing the gold leaves directly.

If the plate or knob becomes electrified by any means the gold leaves will also become charged with like electricity and will repel each other; the distance to which they stand apart will clearly depend in some way on their state of electrification and may be utilized to determine whether the original source of the electrification is strong or weak.

We can perform a number of experiments with the gold-leaf electroscope to illustrate the matters already referred to as well as some of the fundamental laws of the subject.

EXPERIMENT 3. *To determine whether a body is electrified and to shew that electrification is produced by friction.*

Bring a body such as a rod of ebonite or sealing-wax near the knob of an electroscope and observe what happens; if the gold leaves do not diverge the body is not electrified, if they do, the body is charged; in this case discharge the body either

---

[1] For delicate work fused quartz is now sometimes used as the insulator instead of the glass rod.

by drawing it across your hand or by passing it through a gas flame. Then rub the rod with a piece of dry flannel or fur. On again bringing it near the electroscope the leaves diverge, the body is electrified ; if the fur is dry and is fastened to the end of an insulating rod and not held in the hand, it may be possible to shew that it is also electrified by the friction, by removing the ebonite and bringing the fur near, when the leaves again diverge.

This last experiment does not always succeed because the damp hand is a conductor and much of the electrification of the fur has escaped in the handling.

EXPERIMENT 4.   *To shew that there are two kinds of electrification.*

This has already been shewn in Experiment 2.

Rub an ebonite rod with flannel and bring it near an electroscope, the leaves diverge ; remove the ebonite rod and bring near a glass rod which has been rubbed with dry silk, the leaves again diverge. Repeat the experiment with the ebonite rod, but while it is near the electroscope bring up the glass rod, note what happens ; as the glass rod is brought near, the leaves begin to fall together again ; the electrification of the glass rod reduces the divergence due to that of the ebonite rod, the two states of electrification are opposite ; if the glass be brought sufficiently near, the leaves may collapse entirely and then begin to diverge again ; if this is the case on removing the ebonite rod they will diverge still more.

The electrification of the glass rod is said to be vitreous or positive, that on the ebonite resinous or negative.

EXPERIMENT 5.   *To charge an electroscope (a) by conduction, (b) by induction.*

(a) Electrify the ebonite rod by friction with a piece of dry flannel, it will be negatively electrified ; allow it to touch the knob of the electroscope, the leaves diverge and remain divergent when the ebonite is removed, though less widely than when the rod was in contact. The rod has given up part of its negative charge to the electroscope which has thus been negatively electrified by conduction. Discharge the electroscope by touching the knob for a moment with the finger ;

on again bringing up the ebonite the leaves again diverge, shewing that only a part of the charge was in the first instance communicated to the electroscope.

(b) Electrify the ebonite again by friction and bring it near the knob of the electroscope, the leaves diverge. When the ebonite is in position touch the knob for a moment with your finger, the leaves collapse; remove your finger and then remove the ebonite rod; the leaves again diverge; the electroscope is charged but in this case the ebonite has not given up any of its electrification, the electroscope has been charged by induction. Moreover the charge in this case is positive. For we have seen that a negatively charged body induces positive electrification in the near parts of any neighbouring conductor and repels negative electrification to a distance. Thus in the first part of the experiment the knob is positively electrified by induction, the leaves negatively. When the electroscope is touched it becomes electrically part of the earth. The knob is still positively charged but the negative electrification passes to the earth. On removing the hand and then the ebonite this positive electrification is free to spread itself over the gold leaves, which thus diverge.

We may shew experimentally that the final electrification is negative in (a), positive in (b), thus: electrify a glass rod by friction with silk, and bring it near the electroscope. In case (a) it will be noted that the leaves collapse as the glass rod is brought near, they are therefore negatively electrified; in case (b) they diverge still further, they are positively electrified.

EXPERIMENT 6. *To illustrate the difference between conductors and non-conductors and to charge a conductor by friction.*

Charge the electroscope and then touch the knob with various substances held in the hand, taking care that none of them are previously electrified, and note the effects. With some substances the electroscope is practically unaffected, these are insulators; with others the leaves immediately collapse, these are good conductors; with others again there is a slow fall of the leaves, these last substances are bad conductors.

Again take a brass tube on an ebonite handle and holding

the brass in the hand rub it with flannel taking care not to rub the ebonite. On bringing the brass up to an uncharged electroscope no effect is produced, the brass tube is not electrified. Now hold the brass by the ebonite handle and again rub it with the flannel, on bringing it up to the electroscope the leaves diverge, the brass has been electrified by friction, and the electrification has been prevented from escaping by the ebonite handle.

EXPERIMENT 7. *To test whether a charge is positive or negative.*

Charge the electroscope with positive electricity by induction as in Experiment 5. Electrify a glass rod by friction with silk and bring it near the knob of the electroscope. Observe that the leaves diverge further, the positive electrification of the glass repels more positive electricity into the leaves.

Electrify an ebonite rod by friction with flannel and bring it near the knob. At first the divergence of the leaves is decreased, and as the rod is brought nearer, if it be strongly electrified, they collapse entirely, on bringing the rod nearer still they diverge again, but this time with negative electricity.

It is therefore necessary to observe the first indication of the electroscope.

Thus, starting with an electroscope positively charged, if the approach of a body causes still further divergence the body is positively charged, if it causes the leaves to collapse the body is negatively charged.

EXPERIMENT 8. *To shew that both kinds of electricity are produced simultaneously (a) by friction or (b) by induction.*

We have already proved this by means of experiments with the red lead-sulphur powder; in the following experiments the gold-leaf electroscope is used.

(a) Tie a piece of dry flannel or of catskin on to an ebonite rod, and having charged an electroscope positively rub the flannel on a second ebonite rod. On bringing this ebonite near to the electroscope the leaves collapse, it is negatively charged; on removing the ebonite and bringing the flannel near they open more widely than before, the flannel is positive.

The flannel is tied to the ebonite so that it need not be handled, the ebonite is a much better insulator than the flannel.

(b) Obtain two equal metal balls A and B mounted on insulating stands and place them eight or ten centimetres apart. Connect them by a piece of wire held in an insulating handle so that the connexion between the two can be broken without touching either with the hand. Bring a body charged positively up near the ball A. Then remove the connecting wire and examine the electrifications of A and B by bringing each in its turn up to a charged electroscope. In doing this care must be taken to handle only the insulating stands of the balls. It will be found that A is charged negatively, B positively. If the inducing body had been negative then A would be positive, B negative. The two opposite electrifications are produced by induction.

A block of paraffin wax makes a very good insulating stand for this and similar experiments.

EXPERIMENT 9. *To prove there is no electrification inside a hollow closed conductor provided there is no insulated charged body within it.*

For the hollow closed conductor in this experiment, we use a tall metal can 15 to 20 cm. in height, which should be well insulated by being placed on a block of clean paraffin. The can is not completely closed but with the accuracy to which we can work the error due to this is negligible. A small brass ball about a centimetre in diameter, supported either by an ebonite handle or by a silk thread, is also wanted.

Place the conductor on an insulating support and charge the ball either by induction or by the use of some electrical machine (see Section 49). Verify that the ball is electrified by bringing it near to the electroscope. Place the ball inside the conductor and let it touch the side, then lift it out and again bring it near the electroscope, the leaves remain undisturbed, the ball has lost its charge by contact with the interior of the conductor. Repeat this several times, charging the ball and then allowing it to touch the interior of the conductor ; on lifting it out the ball is in all cases completely discharged ; its electrification has passed to the exterior of the conductor. Verify that the exterior of the conductor is electrified either by touching it on the exterior with the ball and then bringing the ball near to the electroscope when the leaves diverge, or by bringing the conductor near the electroscope—if this last method be adopted, the conducting part of the vessel must not be handled.

Thus whenever an electrified conductor is placed inside a hollow conductor and allowed to touch it, the electrification passes at once to the exterior of the hollow conductor, even though that be already charged. There is no electrification inside.

We have supposed throughout this experiment that there is no second electrified body within the hollow conductor but insulated from it. If we had a second charged ball and held it inside the conductor without allowing the two to touch, we should find on repeating the experiment that the first ball was not completely discharged by contact with the conductor.

EXPERIMENT 10. *To prove that there is no electrical force inside a hollow closed conductor provided there is no charged insulated body within.*

Make a cage of fine wire gauze or thin sheet metal to surround the gold-leaf electroscope entirely, if sheet metal is used pierce in it two or three holes through which the gold leaves can be observed. Charge the cage by the aid of an electrical machine or in any other way, the gold leaves remain undisturbed. There is no force inside ; if, however, a charged body be introduced into the cage the leaves diverge and indicate that there is electric force. The metal cage entirely screens the electroscope from all external electrical force. There is no force inside due to external electrification.

We see hence one reason for surrounding the gold leaves with gauze or with strips of tin-foil as described in Section 7.

8. The Proof Plane. A proof plane, Fig. 5, is a useful piece of apparatus by which the state of electrification of the

surface of a body can be examined. It consists of a disc of
metal curved so as to fit the surface of
the body approximately and supported
by an insulating handle. When the
proof plane is placed against the charged
body it becomes practically part of the
surface of that body. The electricity
from the surface under the proof plane
passes to the outer surface of the proof
plane itself. On removing the proof
plane this electricity is removed with
it, and thus the state of electrification
of any portion of the surface can be
examined.

Fig. 5.

9. No electrification within a closed conductor.

The results of the two last Experiments are of great
theoretical importance, and various other experiments have
been devised to illustrate them.

Thus in Fig. 6 *A* represents a sphere suspended by an insulating thread. *B* and *C* are two hollow hemispheres which fit
it closely and have insulating handles attached. Electrify *A*,

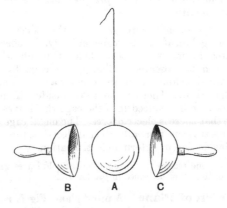

B   A   C

Fig. 6.

B and C being unelectrified. Then fit the two hemispheres together over A, remove them and bring them to the electroscope, they will be found to be electrified. On testing A it will be seen that it has completely lost its electrification.

Faraday's butterfly net experiment is illustrated in Figure 7.

Fig. 7.

A muslin net, a butterfly net attached to a vertical wire, is held in an insulating stand. A silk thread is attached to the vertex of the net, and by pulling this either surface of the muslin can be brought to the outside.

Charge the net with electricity and examine its surfaces by aid of a small proof plane, touching the surfaces in turn and bringing the plane after each contact up to the electroscope. It will be found that the outer surface is charged, the inner uncharged. Now invert the net by pulling the loose end of the thread so that the inner surface becomes the outer. The electricity is on the new outer surface; the surface which was originally uncharged has passed to the outside and has become charged, and *vice versâ*.

In another experiment Faraday constructed a large metal box. He placed inside this box the most delicate electroscopes he possessed and went inside it himself. The outside of the box was then electrified as strongly as possible, but he was not able to discover by the most delicate means in his power any trace of electrification inside.

We have already seen that two bodies similarly electrified repel each other, we have not discussed what the law of force between them is. Now this law can be deduced mathematically from the experimental fact that there is no force inside a hollow closed conductor, and hence we see the theoretical importance of Faraday's experiments.   See Section 26.

CHAPTER II.

ELECTRICITY AS A MEASURABLE QUANTITY.

10. Quantity of Electricity. Hitherto we have spoken in general terms about electrifying a body or charging it with electricity, and we have seen that both kinds of electrification, positive and negative, are produced simultaneously. The experiments moreover have been qualitative, we have not attempted any quantitative measurements. We are now about to describe experiments which shew us that we can look upon an electrical charge as a quantity which like other quantities can be measured in proper units, and that we may speak of charging a body with a definite quantity of electricity just as we speak of pouring a definite quantity of water into a pail. While furthermore, not only are the opposite electricities always produced simultaneously, but they are produced in equal quantities.

EXPERIMENT 11. *To justify the use of the term Quantity of Electricity. Faraday's ice-pail experiments.*

Place a tall narrow metal vessel—the can used in Experiment 9 will do well—on an insulating stand and connect it as in Figure 8 to the electroscope. A second smaller vessel, in Figure 9, of similar shape but with an insulating handle and a number of metal balls fastened to silk threads or insulating handles will be wanted.

Charge one of the metal balls and introduce it into the metal conductor, taking care that the two do not touch. The electroscope leaves diverge, and it may be shewn by a test that their electrification is of the same sign as that of the charged ball. Move the ball about inside the conductor, then provided it is kept some way from the mouth it will be found that the divergence of the leaves remains the same however the position of the ball is altered.

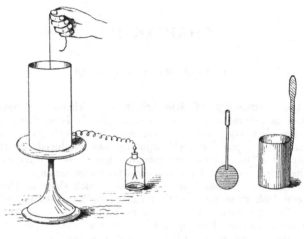

Fig. 8.   Fig. 9.

Introduce now the smaller vessel as well as the ball inside the conductor, taking care to remove previously any charge it may possess. The divergence of the leaves remains the same. Place the ball inside the smaller vessel and let the two touch. Still the leaves remain divergent exactly as before. Remove the ball, it will be found to be unelectrified and the leaves diverge as before. Introduce some uncharged balls carried by insulating threads within the insulated conductor, the leaves remain apart; allow all the conductors to touch the interior of the vessel, this does not affect the leaves; then remove them all, they will be found to be discharged and the leaves remain as before.

Now all these various results are consistent with the statement that originally we put a definite quantity of electricity inside the tall vessel, and that the effect on the electroscope depends solely upon that quantity. By our various operations we have altered the distribution of that quantity, nothing that we have done has changed its amount. By the last operation we transferred the electricity, still unchanged in quantity, from the inside to the outside of the vessel.

By electrifying a conductor we have produced some change in it which remains unaltered in amount as measured by the effect on the gold leaves unless it is allowed to come into contact with another conductor. So long as the first conductor is insulated the charged state remains constant in quantity.

The experiments just described and others which follow were first performed by Faraday using ice-pails for the tall closed conductors, hence the name now given to them.

EXPERIMENT 12. *To shew that the induced charge is equal and opposite to the inducing charge when the conductor in which induction is produced completely surrounds the charged body.*

Connect up the larger "ice-pail" again to the electroscope. Charge one of the balls with positive electricity suppose, and introduce it into the "ice-pail." The leaves of the electroscope diverge with positive electricity. Note their divergence, then touch the ice-pail with the finger, the leaves collapse, the external electrification has escaped to the earth. Remove the finger, and afterwards remove the ball. The leaves again diverge and to the same extent as previously, but their charge is negative.

Thus by introducing the positively electrified ball negative electrification is induced on the inside of the ice-pail and an equal positive charge is driven to the outside and to the leaves. When the pail is touched, this positive charge passes to the earth and the leaves collapse, but on removing the ball the induced negative charge distributes itself over the exterior surface.

Moreover this negative electricity is equal in amount to the original positive charge.

For introduce again the charged ball, the leaves collapse; then allow the ball to touch the inside and remove it. The ball and the "ice-pail" are both discharged, hence the negative charge on the "ice-pail" was equal and opposite to the positive charge on the ball.

EXPERIMENT 13. *To determine if the charges on two bodies are equal.*

Place one ball within the ice-pail and note the divergence of the leaves, remove it and replace it by the second, if the leaves again diverge to an equal extent, the two bodies have equal charges.

Some difficulty is introduced into all these experiments by leakage, it is necessary for success that the insulation should be good and that the operations should be rapidly performed.

With a view to rendering the gold-leaf electroscope more useful for quantitative experiments, a scale is sometimes fixed behind the leaves. The amount of divergence can then be noted on the scale and it can be seen more readily whether the divergence in two different experiments is the same or not.

We may use the ice-pail arrangements for some other experiments.

EXPERIMENT 14. *To prove that equal and opposite quantities of electricity are produced by friction.*

This is practically a repetition of Experiment 4.

Fasten a piece of fur on to an ebonite handle and rub another ebonite rod with the fur, taking care that all the apparatus is originally free from charge.

On introducing either the fur or the ebonite into the "ice pail" the leaves diverge. Introduce the two together. No divergence is observed, the charges on the fur and the ebonite are equal and opposite.

Or again, to prove that the quantities of positive and negative electricity produced by induction are equal and opposite, we may repeat Experiment 12.

That experiment has shewn that the two balls are oppositely charged. Place the two simultaneously within the "ice-pail" no divergence will be observed, hence the quantity within the ice pail is zero, the two charges are equal in amount.

11. Theories of Electrical Action. It appears
from these and similar experiments that it is quite reasonable to speak of a Quantity of Electricity and to suppose
that there are two opposite kinds of electricity which we call
positive and negative. Now we have seen already that two
similarly electrified bodies, or as we may now say two bodies
charged with quantities of similar electricity, repel each other ;
we can observe, and in some cases measure, the force of repulsion between the *bodies*.   But it is quite consistent with, and
will enable us to coordinate, the results of observations to
suppose that the force acts between the like quantities of
electricity. The force observed between the charged bodies
may really be the resultant of the forces between the like
electrifications on those bodies ; the electrical charges are
attached to the bodies, and the action between the charges
may shew itself in a motion of the bodies. We will suppose
therefore that there is a force of repulsion between two like
charges and one of attraction between two unlike charges.

Let us now further suppose that an uncharged body
contains equal quantities of positive and negative electricity ;
these in the undisturbed state exactly neutralize each other,
and the body produces no electrical effects ; if we can in any
way remove some of the negative electricity the body remains
positively charged and *vice versâ*.

12. Explanation of Electrostatic Actions. We
do not know what electricity is, how it is connected with
matter, or what constitutes a state of electrification. We
can however from the above hypotheses give a consistent
explanation of many electrical phenomena.

For example, bring the body near a positively charged
body, the positive electricity in the first body is repelled
and the negative attracted by this charge, thus the body
becomes electrified by induction ; if we can divide it into
two parts as in the case of the two balls in Experiment 12
we obtain two bodies, one charged positively, the other
negatively.

Or again, when electrification is produced by the friction,
say, of silk on glass, we must suppose that the two electricities
are separated by the friction, and that the glass retains more

than its normal share of the positive, the silk more than its
normal share of the negative.

13. Electrical Distribution. Surface Density.

If we are to look upon the electrification of a body as the
distribution of electricity over the surface of the body, we
may ask the question, According to what law does the
electricity distribute itself?   Will there be the same quantity
on each unit of area of the surface or will this quantity vary
from point to point?   The mathematical theory of electricity
gives us an answer to this.   It can be shewn that we may
treat the positive electricity as though it were a fluid free to
move over the surface of the body; we must suppose the
particles of this fluid to repel each other according to a
certain law depending on their distance apart, and calculate
from this what the distribution will be.   The negative electricity must then be treated as a similar fluid of mutually
repulsive particles, and we must further assume that there is
an attractive force between the particles of the two kinds of
fluid.

The two-fluid theory of electricity is based on these
assumptions, and from them, the law of force being known, the
distribution of electricity in a number of actual cases has been
calculated.

But without going into any such elaborate calculations as
would be involved in the above, the fact that in general the
distribution of electricity on a charged conductor is not
uniform but depends on the shape of a conductor and its
position relative to other conductors can be shewn by experiment.

DEFINITION.   *The quantity of electricity on each unit of
area—one square centimetre—of a charged conductor is known
as the* Surface Density *of the distribution.*

It is here implicitly assumed that the distribution is uniform over
each square centimetre.   In the case of a non-uniform distribution we
may define the surface density at any point to be the ratio of the charge
on any small area containing the point to the area when it is taken to be
so small that the distribution over it is uniform.   Cf. the difference
between uniform and variable velocity, *Dynamics*, Section 22.

Now we have seen that a proof plane when applied to a conductor and removed takes with it the charge which occupied the portion of the conductor covered by the proof plane. If then we have a proof plane one square centimetre in area and apply it at different points of the surface the charge removed will in each case measure the surface density about that point. Now the equality of these charges can be determined by the "ice-pail" experiment, and though it is not possible numerically to compare the charges by comparing the divergence of the leaves in the different experiments, yet we may say with certainty that a large divergence implies a large charge and *vice versâ*.

If for the electroscope some form of electrometer, Sections 60, 62, be substituted the charges can actually be measured, and hence the densities at different points can be compared.

In this manner it has been shewn for example that on a sphere the density is uniform, on an elongated conductor it is greatest near the ends; when a conductor is brought near a positively charged body it is found to become negative near the body, positive far away. A line along which there is no electrification may be drawn round the body. Again, it may be shewn that the density is always great near sharp points or corners. These should in most cases therefore be avoided in electrical apparatus. Figure 10 shews in a graphical

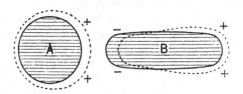

Fig. 10.

manner the distribution of the electricity on two conductors; the distance between the thick line, representing the body, and the dotted line which surrounds it is supposed to represent the density; when the dotted line lies inside the body the charge is negative, when it is outside the charge is positive.

The theory we have just been describing is often spoken of as the two-fluid theory of electricity; if we adopt it provisionally we must guard ourselves against looking upon electricity as a fluid. All that the theory states is that an electric charge distributes itself over the surface which separates a conductor from the insulating material about it, according to the same laws as a fluid consisting of mutually repulsive particles would do were it free to move over that surface. The question whether the electric charge resides on the surface of the conductor or on that of the insulating medium by which it is surrounded is one of great importance, to which we shall return later. See Section 46.

14. Electrical Pressure or Potential. Consider two insulated conductors, A and B, two metal balls suppose, supported by silk threads. Let A be charged and B uncharged. Allow them to touch and then separate them. We can easily shew by experiment that electricity has passed from A to B. Let us ask ourselves the question why has it so passed, and what limits or regulates the flow?

Or, again, take two charged bodies and compare the charges on the two by placing each in turn in the "ice-pail," and observing the effect. Allow them to touch and separate them. On again comparing their charges it will probably be found that the charge on each has altered though the total charge remains the same. Electricity has again passed from the one ball to the other and we may ask the same question.

The flow depends to some extent on the charge on each ball but not on it alone, for we can easily arrange the experiment so that the ball with the larger charge gains electricity and that with the smaller charge loses it.

We can obtain an answer to our question most readily by considering cognate problems in other sciences.

Thus consider two reservoirs containing water, A and B, Figure 11, connected by a tube having a stopcock, and suppose the level of the water in A is higher than that in B. On opening the stopcock the water runs from A to B. The lower reservoir B may contain the most water, the flow is not regulated by this but by the difference in level of the water surfaces in the two, or, and this comes to the same thing, by the difference of pressure between the two ends of the pipe. If we suppose, in order to get rid of the effect of the weight of the water in the pipe, that the pipe is horizontal the water will flow through the pipe from the end at which the pressure is greatest to that at which it is least.

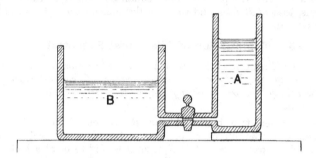

Fig. 11.

Or, again, take two gas-bags filled with compressed air; connect them by an india-rubber tube closed with a stopcock. If the pressure of the air is the same in the two, on opening the stopcock there is no flow. If the pressure be different the flow is from the bag at high pressure to that at low pressure.

The Science of Heat affords another illustration[1]. When two hot bodies are put into thermal communication heat flows from the body at a high temperature to that at a low temperature.

Moreover in these two cases the flow of fluid continues until the pressures in the two reservoirs are equalized, while the flow of heat continues until the temperatures of the two bodies become the same.

Temperature is defined to be the condition of a body on which its power of communicating heat to or receiving heat from other bodies depends.

Corresponding to Pressure in Hydrostatics and to Temperature in the Science of Heat we have in dealing with Electricity to consider the **Electrical Pressure or Electrical Potential,** as it is called.

[1] Glazebrook, *Heat*, § 11.

Electricity passes from a body A to a body B because the electrical potential of A is higher than that of B and the flow goes on until the potentials are equalized: the potential of A falls, that of B rises, and the flow depends on the difference of potential.

15. Explanation of Electrical Potential.

The **Electrical Potential** *of a body measures the condition of the body on which its power of communicating Electricity to, or receiving Electricity from, other bodies depends.*

If when two bodies A and B are put into electrical connexion the direction of the electric force is such that electricity passes from A to B, then A is said to be at a higher potential than B. Two bodies A and B have the same potential if when they are put into electrical communication there is no transference of electricity between them.

These statements should be compared with the corresponding Definition of Temperature[1].

It must also be noted that the above statement does not give us a means of measuring Electrical Potential, it gives a name to a quality of an electrified body on which certain important properties depend. One important consequence of the statement is the following.

PROPOSITION 1. *All points on a conductor in a state of electrical equilibrium are at the same potential.*

For consider two such points A and B, Figure 12, if there be any difference of potential between them electricity will pass until this difference is neutralized, the condition that the electricity should be in equilibrium on the conductor is that the potential should be the same at all points.

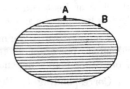

Fig. 12.

In the case of the Earth, which is a large conductor, differences of potential do exist between different points, and in consequence

[1] *Heat*, Section 11.

currents, earth currents, are continually passing, but these are small, and we may treat the potential of the Earth for most purposes in the neighbourhood of any point of observation as being uniform.

16. Zero of Potential. The Earth is so large that any charge we can give it is insufficient to change its potential appreciably, just as the ocean is so big that the rain it receives cannot appreciably affect its level. Just then as we take as our zero level from which to measure heights the mean level of the sea and treat it as fixed, so we take as our zero, from which to measure potential or electrical level, the potential of the Earth.

Thus the potential of the Earth is the zero of electrical potential.

If positive electricity flows from a body to the Earth that body has a positive potential; if positive electricity flows from the Earth to the body the body has a negative potential.

17. Analogy between Pressure, Temperature, and Potential. We have seen that there is a certain analogy between potential and pressure or temperature; there is however one important point of difference to be noted.

The pressure in a reservoir of gas does not depend in any way on the pressure in neighbouring reservoirs; if we have a number of independent vessels filled with gas we may alter the pressure in any one without affecting that of the rest.

Again, in the case of heat, if we could prevent radiation, the temperature of a number of isolated bodies would not be altered by bringing a hot body near; in reality in consequence of radiation heat is transferred in such a case and the temperatures of all the bodies are *gradually* changed.

In the case of electricity however the potential of any conductor depends on that of all the other conductors in its neighbourhood; if we bring a charged body near a number of other charged bodies the potential of each body in the system is very rapidly, practically immediately, altered, the change takes place not by a slow gradual process like the absorption of radiation but at once.

The facts of electrical induction shew that this is the case. On bringing a positively charged body near an insulated conductor there is a separation of the electricities on that conductor, and on connecting it to earth positive electricity passes away, hence before the connexion the insulated conductor had a positive potential. The presence of the positively charged body in its neighbourhood immediately[1] raised its potential.

18. Importance of the Insulating Medium. We have seen that there is a certain analogy between a number of reservoirs filled with compressed air and a number of electrical conductors, but the analogy does not carry us far. We may make it more complete thus.

Imagine a lump of india-rubber or some other jelly-like elastic substance with a number of cavities in it. Suppose that each of these cavities has a pressure-gauge attached and is filled with water. Adjust the quantity of water in each until the gauges all read alike. The cavities represent a number of conductors and the india-rubber the insulating medium between them. Increase the pressure in one of the cavities by pumping more water in or otherwise. The increased pressure will be resisted by the elasticity of the medium, and the pressure everywhere throughout the medium will be increased in the substance of the elastic medium as well as in the cavities; the gauges will all read differently, the increase of pressure in any cavity will depend on its size and its position as well as on the elastic properties of the medium.

In electrical language one of the conductors has been charged, its potential has been raised; this produces a change of potential everywhere; in the conductor itself, in the surrounding medium or dielectric, and in neighbouring conductors, this change depends on the nature of the medium as well as on the size and position of the conductors.

Just as the increase of pressure in a cavity in the india-rubber sets up a state of stress throughout the mass so we may look upon the electrification of a conductor as the pro-

---

[1] It is probable that time is required for the change but the time is extremely short.

duction of a state of stress in the insulating medium round
the conductor; where the continuity of the medium is broken
by the presence of other conductors this state of stress
manifests itself in the phenomenon we call the electrification
of the conductors. From this point of view to produce an
electric charge at any point is to throw the insulating medium
round the point into a state of stress, and the electric
attractions and repulsions we have observed are the consequences of this stress. The insulating medium and not the
conductor is the important factor; it has the power of
resisting the stress and possesses what we may call electric
rigidity; in the conductor this power is wanting, the stress
breaks down as soon as it is applied, all points in the conductor
are at the same electric pressure or potential.

This is the more modern view of electricity and electric
force, and in it the importance of the dielectric is clear.

This view is due to Faraday and was developed by Clerk-Maxwell.
We might carry the analogy further but it is hardly necessary to do this
at present.

19. Equipotential Surfaces. A positively charged
conductor produces electric potential at all points in its
neighbourhood, while all points on the conductor are at the
same potential. The surface of the conductor is said to be an
equipotential or level surface. Consider now one such body,

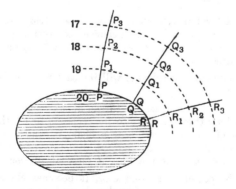

Fig. 13.

Figure 13, the potential falls off as we recede from its surface. Starting outwards from the body along any line PP_1P_2, etc. we can find a series of points P_1, P_2, such that the difference of potential between any two consecutive points is unity. Thus if the potential at P be 20, the potential at P_1 etc. will be 19, 18, 17, etc.

We can do the same for all points Q, R, S, etc., on the surface of the conductor; we thus find a number of points, Q_1, R_1, S_1, etc. all at potential 19; if we suppose all these points joined we have an imaginary surface surrounding the body at all points, of which the potential is 19; this is called an equipotential surface.

Similarly by joining P_2, Q_2, R_2, etc., we get a third equipotential surface at potential 18, and so on; these surfaces surround the conductor like the coats of an onion.

DEFINITION. *An* **Equipotential** or **Level Surface** *is a surface at all points of which the potential has the same value.*

We have described the drawing of an equipotential surface for a single conductor; if we have a number of conductors we can still conceive of corresponding equipotential surfaces; in cases in which it is possible to draw these, the problem of determining the distribution of electric force throughout the field can be solved.

20. Lines of Force. There is electric force at any point in the neighbourhood of a charged conductor or system of charged conductors, and at each point this force has its own definite direction.

The following experiment will indicate in a rough manner the direction.

Take a short bit of cotton thread about a centimetre long and tie a long thin silk fibre to its centre. Bring the cotton by means of the silk fibre near a charged conductor, it will set in a definite position, and the direction of its length is approximately that of the line of force at its centre.

For the cotton is a conductor, it becomes electrified by induction when brought near the charged system, becoming positive at one end, negative at the other; the positive end is pushed in one direction by the electric force, the negative end is pulled in the opposite direction; the equilibrium position will be found when the lines of action of these two forces coincide, that is when the length of the bit of cotton lies along the line of action of the force. In this manner the fact that there is electric force at all points of the field can be shewn, and its direction roughly indicated.

Now let AB, Fig. 14, be one position of the test thread. Suppose the thread moved until A comes into the position B

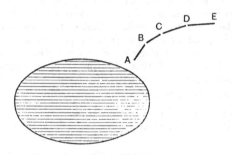

Fig. 14.

it will take up another position BC, then move it again till the end originally at A comes to C, we can find a third position CD and so on. Each of these short lines AB, BC, etc., indicates the direction of the electric force at its centre; if the thread AB is very small these short bits make up a continuous line, usually curved, and this line has the property that its direction at each point of its length gives the direction of the resultant electric force at that point. Such a line is called a line of force.

DEFINITION. *A* **Line of Force** *in an electric field is a line such that its direction at each point of its length gives the direction of the resultant electric force at that point.*

The method of tracing the lines of force just described can not be carried out satisfactorily in practice, although as we shall see, under Magnetism, a similar method is quite practicable and of great value for tracing lines of magnetic force; the experiment, however, although not suited to give accurate results, is sufficient to shew the presence of electric force throughout the field and to give a general indication of its direction, shewing that it acts along lines which are usually curved.

21. Lines of Force and Equipotential Surfaces.

If we trace a line of force we find that it begins and ends on a charged surface, passing in all cases in the direction in which the potential falls, so that if the potential at one point A on a line of force is above that at another point B the direction of the line is from A to B. The electrification at one end of the line will be found to be positive, that at the other end of the line negative, and the line goes from the positive to the negative. Moreover, it will be clear from the direction in which the test thread sets itself, that the line of force meets the surfaces which terminate it perpendicularly.

This leads us to two important propositions:

PROPOSITION 2. *The two extremities of a line of force rest on oppositely electrified surfaces.* This is verified by direct experiment.

If we connect these two surfaces by a conducting wire the extremities of the line of force may be supposed to move up together along the wire, and the line to close down on the wire.

PROPOSITION 3. *The direction of a line of force is perpendicular to that of all the equipotential surfaces which it meets.*

So far as concerns the two surfaces on which the line of force ends, and which are of course equipotential, this also is verified by experiment. A theoretical proof applying to all equipotential surfaces intersected by the line of force may be given thus:

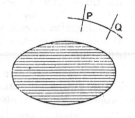

Fig. 15.

Let P, Fig. 15, be any point on an equipotential surface, and Q any other point on the surface adjacent to P. Then since P and Q are at the same potential there is no force tending to move electricity from P to Q. The electrical force at P therefore must be at right angles to PQ. Hence its direction is perpendicular to the surface, in other words, a line of force is perpendicular to any equipotential surface which it meets.

22. Forms of Lines of Force. The student will be helped in this view of electrical action if he can picture to himself the forms of the lines of force in some simple cases. The simplest case is that of a positively charged sphere placed in a large room, the lines of force for such a system, Fig. 16,

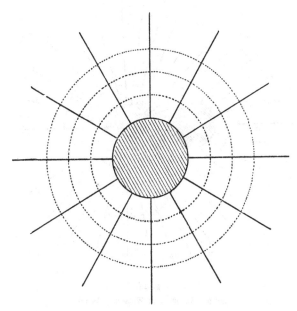

Fig. 16.

are straight lines radiating outwards from the centre of the sphere, the equipotential surfaces shewn by the dotted lines are spheres concentric with the charged sphere; these results follow at once from the symmetry of the system. The walls of the room are conductors and the lines of force end on them. Each line terminates in a negative charge equal to the positive charge at the point from which it starts; the total negative charge on the walls is equal in amount to the positive charge on the ball.

Suppose now that another uncharged and insulated conducting ball is brought into the room; some of the lines of force converge on to this ball, see Fig. 17, meeting it, as

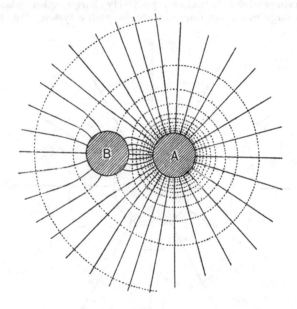

Fig. 17.
By permission, from Watson's *Physics.*

shewn, at right angles. These lines end in a negative charge, so that the side of the ball nearest the charged conductor becomes negatively charged. But the total charge on the ball is zero, hence the opposite side of the ball acquires a positive charge and from this side lines of force equal in number to those entering the ball, start and travel towards the walls of the room. The dotted lines again represent the equipotential surfaces.

In Fig. 18 is shewn the result after the second ball has been connected by a conducting wire to the walls.

The ball has thus become electrically a part of the walls; the lines of force between the ball and the wall have their two ends on the same conductor—this is impossible; the

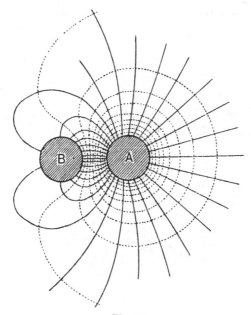

Fig. 18.
By permission, from Watson's *Physics*.

oppositely electrified ends move up together; the lines close in on the wire and collapse, and we are left with the distribution figured.

The tension along the lines of force in both these cases tends to draw the two conductors together.

The distribution on two similarly charged spheres is shewn in Fig. 19.

The lines run from the spheres to the wall; those which

start from either sphere towards the other are repelled as it were by the corresponding lines from the second sphere. The point P between the spheres on the line joining their centres is a point at which there is no force; the line through this point perpendicular to the axial line is a line of force.

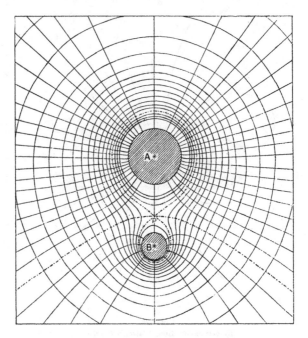

Fig. 19.

The tension along the lines between the spheres and the walls draws the spheres apart; they apparently repel each other.

Fig. 20 gives two oppositely charged spheres.

A number of the lines from the positive sphere bend round and converge on the negative sphere; the negative sphere is at a lower potential than the walls. Hence lines pass from the walls to it while other lines pass from the positive sphere to the walls. The tension along the lines between the two spheres draws them together; they attract each other. The *P* is again a point of zero force.

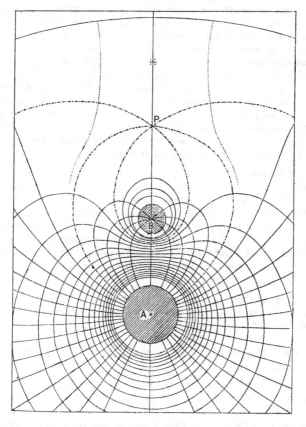

Fig. 20.

23. The Field of Force. We have thus another method of picturing the field due to a charged conductor; we may think of it as permeated by the lines of force due to that conductor, lines, that is, along which the electric force due to the charge on the conductor acts. If we start outwards, along one of these lines, and follow it up, we shall find it will end somewhere in a negative charge on some other conductor; between these two charges there is apparently a force of attraction. If we wish to examine further the mechanism by which the force of attraction is produced we may find it in the insulating medium between the two. We have already seen that the electrification of a conductor may be the manner in which we recognise a state of stress set up in the dielectric round the conductor. Let us suppose that this state of stress is such as to tend to cause the dielectric to contract everywhere along the lines of force and to expand everywhere at right angles to them. Such a tendency, if of the proper amount, would, it may be shewn, produce exactly the forces in the field which experiment shews to exist.

If we are considering the force between two small charged bodies we may either speak of the direct attraction between the two, or we may picture to ourselves the lines of force which join them, and think of the dielectric as tending to shrink up along these lines and so draw the two bodies together; each line of force has a tendency to contract or close up on itself like a stretched elastic thread. Like the thread also, it tends to swell laterally at the same time as it shrinks longitudinally.

This method of considering electrical action is due to Faraday; according to it as we see everything depends on the properties of the dielectric.

24. Mechanical Illustrations. We may illustrate the matter further thus :

Suppose we have two balls connected by a number of fine elastic threads ; we may imagine the threads to be carried over pullies, or otherwise guided in some way so that they are not all straight, but pass from the one body to the other in curved lines.

Let the bodies be pulled slightly apart so as to stretch these threads. Then to an observer who was not conscious of the threads it would appear that there was an attractive force acting between the bodies across the empty space which separates them ; one who could recognise the existence and functions of the threads would see that the attractive force was the manifestation of the stress set up by stretching them.

Or again, to give one more illustration, we might look upon a line of force as a string of little cells each filled with liquid connecting the two electrified bodies like beads upon a thread. Now suppose that these cells can be made to spin rapidly about the line of force as an axis. Each cell will tend to flatten itself along the axis and to expand in directions perpendicular to the axis. The lines of force will tend to contract along their length and to squeeze each other outwards in other directions. There will be a tension in the dielectric along the lines of force, and a pressure perpendicular to them. The stress which we know exists in an electric field might be produced thus.

25. Theory of Potential applied to Electrical Attraction. In the earlier Sections we have already described and explained various experimental results on the hypothesis of mutual attractions or repulsions between electrified bodies. Let us consider how some of these are modified if we introduce the notions of potential and lines of force. Take for example the phenomena of induction when a positively charged sphere is brought near an insulated conductor.

We know that (i) the sphere produces electrical potential or pressure at all points in its neighbourhood and that this potential falls off as we move away from the sphere[1].

(ii) All points on a conductor in electrical equilibrium are at the same potential.

(iii) Positive electricity passes along a conductor from points at high to points at low potential.

In Fig. 21 let the consecutive circles indicate the equipotential lines due to the sphere before the insulated conductor is brought near, and let the numerals 10, 9, 8, etc. represent

[1] This last law has not been directly proved up to this point, but see Section 26.

the potentials of these surfaces. On bringing the conductor
near it cuts a number of these surfaces; if there were no

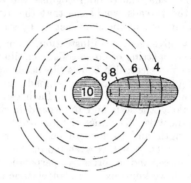

Fig. 21.

change in potential the potential of one point on the conductor would be 8, of another 7, of another 6, and so on.
This however is impossible.

Positive electricity passes from one end of the conductor,
lowering its potential, to the other which has its potential

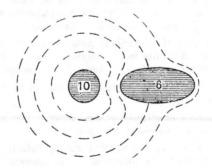

Fig. 22.

raised; the result is that the first end is left negatively
charged, while the second acquires positive electrification.

This transference goes on until the whole conductor is reduced to the same potential and the equipotential surfaces due to the sphere are distorted in the process until we reach the condition represented in Fig. 22.

Lines of force now pass from this sphere to the conductor and the tension along these is shewn in the attraction observed between the two.

It should be noticed in this case that the potential of the conductor is positive though it is negatively charged at one end.

The leaves of a gold-leaf electroscope when charged positively are at a higher potential than the walls of the case; these if coated with gauze or tinfoil in connexion with the table are at zero potential. Lines of force therefore pass from the leaves to the walls and the tension along these lines causes the leaves to stand apart.

If we suppose the electroscope charged by induction from a positively charged body we may represent the condition of affairs thus. Lines pass from the positively charged body to the disc or knob which becomes negatively electrified; a positive charge is thus driven to the leaves, and lines of force pass from them to the cage. The tension along these lines is sufficient to draw the leaves apart; hence the divergence which is observed.

CHAPTER III.

MEASUREMENT OF ELECTRIC FORCE AND POTENTIAL.

26. Law of Force. Up to the present it has not been necessary to consider what is the force between two electrified bodies or how the potential in the neighbourhood of a charged body is to be measured.

We have learnt that two bodies charged with like electricities repel each other, and a very little observation is sufficient to shew that the force increases as the distance between the bodies is decreased; thus the divergence of the leaves of an electroscope caused by the presence of a charged body increases as the body approaches the electroscope.

We must now proceed to consider according to what law the force decreases.

In the first place, as already stated, we have no means of observing the action between two *charges of electricity*; all we can do is to measure the force between *two bodies* charged with electricity. This measurement was first made by Coulomb by aid of a piece of apparatus called a torsion balance[1], which is described in Section 59, and it follows as a result of the experiments that when two similarly charged small bodies are at a distance apart which is great compared

---

[1] Among the many interesting pieces of historical apparatus in the Paris Exhibition of 1900 was the original torsion balance used by Coulomb.

with their dimensions, there is a repulsion between them which is proportional to the product of their charges and inversely proportional to the square of the distance between them.   Hence Coulomb's law may be expressed thus:

LAW OF ELECTRIC FORCE.   *Consider two small bodies charged with quantities e, e′ of electricity and placed at a distance r apart, the distance r being great compared with the size of either body.   Then there is a force of repulsion between these bodies which is proportional to* ee'/r^2.

Now we suppose that this law which is shewn to be true within the limits of experimental error for two small bodies would be accurately true for two electrified points.   And in the mathematical theory of electricity we treat it as the law of force between the particles of electricity, and our real proof of the law lies in the fact that the results of complicated experiments can be determined by calculation on the assumption of the truth of this law and are found to agree with observation when the experiments are performed.

In particular Faraday's result that there is no electric force inside a hollow closed conductor which does not contain an insulated charged conductor is consistent with this law of force, and with this law only, and it is this fact which forms the real basis of our belief in the law.

27. Effect of the Dielectric; Inductive Capacity.
We have said already that according to the modern view of electricity the action between electrified bodies depends in great measure on the dielectric or insulating medium which separates them, and it is found that the force between two charged bodies can be changed by changing this medium without altering the charge or position of either body.   Experiments to illustrate this are described in Section 42.

Suppose, for example, that it is possible to measure the force between two bodies when separated by air and that without making any other change the air is replaced by paraffin oil. Then it will be found that the force is about half[1] as great as it was.

----

[1] To be accurate it is reduced in the ratio 1 to 1·92.

If for paraffin oil we substitute sulphur then the force will be reduced in the ratio 1 to 3·97, it will be about one-fourth of what it was in air. This result is expressed by introducing into the law of force a factor to express the action of the dielectric. This factor is known as the inductive capacity of the dielectric. If we denote it by K and if F be the force between two particles with charges e, e' at a distance r apart, then we have as a more complete statement of the law the result that

$$F = \frac{1}{K} \frac{ee'}{r^2}.$$

Since in a very large number of experiments air is the dielectric medium, it is found convenient to assume that its inductive capacity is unity, and to measure the inductive capacities of other media in terms of this. This is known as the electrostatic system of measurement.

The ratio of the inductive capacity of a medium to the inductive capacity of air is known as the **Specific Inductive Capacity** of the medium.

On the electrostatic system, since the inductive capacity of air is assumed to be unity, it follows that the numerical measures of the inductive capacity of a medium and of its specific inductive capacity are the same.

In considering this subject the student should compare the definitions of density and specific gravity; when water is the standard the numerical measures of these two quantities are the same.

28. Units of Measurement. We have not up to the present defined in any way the units in which the quantities are measured. As usual we employ the c.g.s. system; the unit of distance is the centimetre, the unit of force the dyne; the unit quantity of electricity requires definition.

We have the law that, if the force between two quantities e, e' placed in air at a distance r centimetres apart is F dynes, then

$$F = \frac{ee'}{r^2} \text{ dynes.}$$

Now let the quantities e, e' be equal and let them be such that when placed at a distance of one centimetre apart, the force of repulsion is 1 dyne.

Then in the above expression we have $e = e'$, $F = 1$ dyne, $r = 1$ cm.

Hence
$$1 = \frac{e^2}{r^2}.$$

$$\therefore e^2 = 1, \quad e = \pm 1.$$

Thus in this case e must be the unit quantity of electricity.

Or in other words,

Definition. *The* Unit Quantity of Electricity *is that quantity which placed at a distance of 1 centimetre in air from an equal and similar quantity repels it with a force of 1 dyne.*

On this assumption then as to the unit quantity we have the law that on the electrostatic system in air

$$F = \frac{ee'}{r^2} \text{ dynes};$$

and in a medium of specific inductive capacity K,

$$F = \frac{1}{K}\frac{ee'}{r^2} \text{ dynes}.$$

Thus if we call F_1 the force between the two quantities e, e' when at a distance r centimetres apart in air, and F_2 the force between the same two quantities when in a medium of specific inductive capacity K,

then clearly $\quad F_1 = ee'/r^2, \quad F_2 = ee'/Kr^2.$

Hence $\qquad F_1 = KF_2$ or $K = F_1/F_2.$

Thus the specific inductive capacity of a dielectric is the ratio of the force between two charges in air, to the force between the same two charges when in the dielectric when the distance between the two charges remains unaltered, which explains the definition given above.

29. Resultant Electric Force at a point; Electric Intensity.   At any point in the neighbourhood of a charged conductor there is electrical force; if we place at that point a small body charged with a quantity e' of electricity, then this charge is repelled or attracted by the charge on the conductor, and we can in theory at least calculate the amount of the force by calculating the force exerted by each element of the charge on the conductor, and finding the resultant of these. This resultant force will be proportional to the charge on the small conductor, for each component force is so; and we may denote it by Re'. Thus R is the force which would be exerted on a small body at the point in question when carrying a unit positive charge.

The quantity R is known either as the resultant electric force or as the electric intensity at the point.

It must be noticed however that if we were actually to bring a small body so charged near to the conductor, the distribution of electrification would be disturbed; in calculating the force theoretically we suppose this disturbance not to occur, so that the resultant force could not really be measured experimentally by bringing up a small charged body and determining the force acting on it.

We are thus led to the following:

DEFINITION. *The* **Resultant Electric Force** *or the* **Electric Intensity** **at a point** *is the force acting on a small body charged with a unit positive charge when placed at the point, the electrification of the rest of the system being supposed undisturbed by the presence of this unit charge.*

If the electrical field be due to a charge e concentrated at a distance r from the point at which the resultant force is to be measured, we know that the force between two small bodies carrying charges e and e' respectively is ee'/r^2. The electric intensity is the force on a small body carrying unit charge. Hence we see, by putting e' equal to unity, that in this case the electric intensity is e/r^2.

Thus the **Resultant Electric Force** or **Electric Intensity** at any point due to a charge e concentrated at a distance of r centimetres from that point is e/r^2.

As an illustration of the magnitude of the electrostatic unit we may note, that if two equal small conducting spheres each $\frac{1}{2}$ gramme in weight are suspended from the same point by two silk fibres of inappreciable weight 100 centimetres in length and equally electrified until they separate to a distance of 10 centimetres, then the charge on each must be about 22 units.

Examples. (1) *Calculate the force between two small spheres charged with 10 and 15 units respectively placed at a distance of 50 cm. apart.*

$$\text{Force} = \frac{10 \times 15}{50^2} \text{ dyne} = \frac{3}{50} \text{ dyne}.$$

(2) *The force between the two spheres in Example* (1) *is known to be ·02 dyne. What is the specific inductive capacity of the medium?*

The specific inductive capacity is equal to the ratio of the force in air to the force in the medium. The force in air = 3/50 = ·06 dyne,

$$\therefore \text{ specific inductive capacity} = \frac{\cdot 06}{\cdot 02} = 3.$$

(3) *How far apart must two small spheres each charged with 100 units be placed in air in order to repel each other with a force equal to the weight of one gramme?*

The weight of one gramme = 981 dynes. Let the distance be r centimetres.

Then
$$981 = \frac{100 \times 100}{r^2}.$$

$$\therefore r = 100/\sqrt{981} = 100/31\cdot32 = 3\cdot19 \text{ cm.}$$

30. Measure of Potential.

In Section 15 we have explained in a general manner the meaning of the term Electrical Potential, but the account there given will not enable us to measure potential. Now we have seen already that there is a close analogy between the relation of the flow of heat and of difference of temperature on the one hand, and that of electrical force and difference of potential on the other. Let us carry this rather further.

We have seen (Glazebrook, *Heat*, § 141) that if t and t' are the temperatures of two points d cm. apart, and if the temperature gradient be uniform between these points, then the flow of heat anywhere between the points is constant, and we have the result that flow of heat is proportional to $(t - t')/d$.

Take now a case in which the potentials at two points d centimetres apart are V and V', and the potential gradient is uniform. Then our analogy would lead to this result that the electrical force anywhere between the points is constant and is proportional to $(V - V')/d$.

Let us consider the result of assuming this law to hold, and suppose that the electrical force between the points is R. Then it follows that R is proportional to $(V - V')/d$, and if we choose our units of measurement aright we can put

$$R = \frac{(V - V')}{d}.$$

Hence $\qquad\qquad V - V' = Rd.$

Now R is the force on a small body carrying a unit positive charge, and Rd is the work done in moving the body a distance d against the force.

Thus we shall obtain a result quite consistent with what precedes if we adopt the following

DEFINITION. *The* **Difference of Potential** *between two points A and B is the work done in moving a small body carrying a unit positive charge, from B to A, against the electrical force, the electrification of the rest of the system being supposed undisturbed by the presence of this unit charge.*

We have been led to this definition by considering the case of a uniform force and the analogy with the flow of heat ; by adopting it in general we are enabled to measure in theory at least the electrical potential at any point of the electric field due to a charged system. Practical methods of measuring potential will be given later.

It should be noticed that the definition gives us a means of measuring difference of potential, it does not enable us to say without further explanation what the potential at a point is ; if however we go sufficiently far from our charged system the force due to it will be very small, so small that we may neglect it, no work will be done in moving our unit charge so long as we keep far enough away ; all points therefore at a sufficient distance from our charged system are at the same potential, and we may take this as the zero of potential for these purposes ; these points are said to be beyond the boundaries of the electrical field of the charged system, and in this case the potential at the point might be defined as the work necessary to bring a small body carrying a unit positive charge from beyond the boundaries of the field up to the point.

For some other purposes again it is convenient to treat the potential of the earth as the zero of potential; as we have said already, in consequence of the size of the earth, we cannot alter its potential by any finite charge which we can give it; in all cases in which we speak of the potential of a body as being V we mean that it is V units of potential above some body taken as the zero, and that V units of work are necessary to move a small body carrying unit positive charge from that zero of potential to the body in question, the electrification of the rest of the system being supposed unaffected by the motion.

31.   Unit of Potential.   It is clear that since potential is measured by work done, there will be unit difference of potential between two points when the work done in moving unit charge is the unit of work, the unit of work is called as we know[1] the erg, and is the work done on a particle which is moved one centimetre against a force of one dyne.

DEFINITION.   *There is* **Unit Difference of Potential** *between two points when one erg of work is done in moving a small body carrying unit positive charge from one point to the other, the electrification of the rest of the system being supposed unaffected by the motion.*

In dealing with the subject of potential we might have commenced with the Definition given in Section 30, thus introducing the potential at a point as a quantity which we can measure, and which has an important place in electrical theory. Such a course would have tended to greater precision. Since however the general idea of potential can be utilized as has been done in Section 25, without a strict definition of its measure, the course adopted has seemed the better in the hope that some students who have a difficulty in grasping the idea of work done may yet not be prevented from using this conception of potential as a quality of a body on which the direction of the electric flow depends.

32.   Calculation of Potential.   As we have already said the difference of potential between any two points can be calculated in theory when we know the field, let us exemplify this, calculating the potential at a point P, Figure 23, due to a charge Q at a point O.

---

[1] *Dynamics*, Section 124.

Join OP and let OP equal r.

Produce OP to a great distance—beyond the limits of the

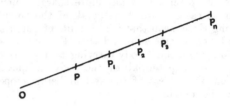

Fig. 23.

field—and let P_1, P_2 be two points on OP at distances r_1 and r_2 respectively from O. Then the force at P_1 is $Q/r_1{}^2$, that at P_2 is $Q/r_2{}^2$, and if P_1 is very close to P_2, $r_1{}^2$ is very nearly equal to $r_2{}^2$, each may be put equal to $r_1 r_2$, thus ultimately we may treat these forces as equal, and equal to $Q/r_1 r_2$.

Thus when the distance $P_1 P_2$ which is equal to $r_2 - r_1$ is sufficiently small, the work done in bringing unit positive charge from P_2 to P_1, being equal to the product of the force and the displacement, is $Q(r_2 - r_1)/r_1 r_2$, and this work is the difference of potential between P_1 and P_2.

Hence
$$V_1 - V_2 = \frac{Q(r_2 - r_1)}{r_1 r_2} = \frac{Q}{r_1} - \frac{Q}{r_2}.$$

By taking another point P_3 beyond P_2, but very near to it, we can get
$$V_2 - V_3 = \frac{Q}{r_2} - \frac{Q}{r_3}.$$

Thus if we go to a point P_n at a distance r_n from O we have the equations
$$V_1 - V_2 = \frac{Q}{r_1} - \frac{Q}{r_2},$$
$$V_2 - V_3 = \frac{Q}{r_2} - \frac{Q}{r_3},$$
$$\dotsc\dotsc\dotsc\dotsc$$
$$V_{n-1} - V_n = \frac{Q}{r_{n-1}} - \frac{Q}{r_n}.$$

Hence by adding these we have

$$V_1 - V_n = \frac{Q}{r_1} - \frac{Q}{r_n}.$$

But P_1 and P_n are any two points on the line OP. Denoting them by P and P' and their potentials by V and V' we have

$$V - V' = \frac{Q}{r} - \frac{Q}{r'}.$$

If we suppose P' to be at a very long distance away we may put $V' = 0$, and since r' is infinite Q/r' is zero. Thus we find

$$V = \frac{Q}{r}.$$

We have thus obtained an expression for the potential due to a charge Q concentrated at a point.

We may notice that the potential due to such a charge at any point is inversely proportional to the distance of the point from the charge, we have already seen that the resultant electrical force or the electric intensity is inversely proportional to the square of the distance.

33. Equipotential Surface. We may use the above result to draw the equipotential surfaces and lines of force due to such a charge.

Thus, since the value of the potential at a distance r is Q/r, for all points at which the potential has a definite constant value r must be the same; such points therefore are at the same distance from the charge, that is they lie on a sphere. Hence the surfaces of equal potential are spheres; the lines of force are the radii of the spheres as shewn in Figure 16.

Or again take the case of an electrified conducting plane of considerable area. The plane itself is equipotential, and the equipotential surfaces except near the edges will be parallel planes while the lines of force will be straight lines at right angles to the planes.

34. Potential of a Charged Conductor. We have already defined the potential at a point as the work done in bringing a small body carrying unit positive charge up to the point; now the point may be on a conductor, and if this is so the potential at the point will be the potential of the conductor. Work has to be done to bring the unit positive charge up to any point on the conductor. This work, if we suppose the electrification of the conductor unaffected by the presence of the unit charge, will be the same whatever point of the conductor be approached, and the amount of work done measures the potential of the conductor.

35. Relation of Charge and Potential. The charge on a conductor will be connected with its potential; the potential of a conductor, however, as we have seen does not depend solely on its charge, but also on the condition of the other conductors, if there be any, in its neighbourhood: let us suppose that all other conductors near are maintained at zero potential by being connected to the earth, then there is a simple relation between the charge and potential of the one insulated conductor. For consider such a charged body; the force at any point in its field is, as we have seen, the resultant of the forces arising from its electrical charge together with the forces which arise from the induced charges on the neighbouring uninsulated bodies.

If the charge at each point of the surface of the insulated conductor be doubled, each of these induced charges will be doubled, and hence each of the various component forces will be doubled. Thus the resultant force at each point of the field is doubled; hence the work done in bringing up a unit charge to that point will be doubled, that is, the potential at each point of the field is doubled.

Hence by doubling the charge on the insulated conductor we have doubled the potential everywhere.

Thus if all other conductors in its field are at zero potential, the potential on any one insulated conductor is proportional to its charge.

36. Capacity of a Conductor.

DEFINITION. *The ratio of the charge on an insulated conductor to the potential of that conductor, when all neighbouring* conductors are at zero potential, is found to be constant and is called the **Capacity** of the conductor.

If C be the capacity of a conductor, Q its charge, and V its potential, other conductors being at zero potential, then C is a constant for the conductor and we have the relation

$$C = \frac{Q}{V},$$

or as we may put it

$$Q = CV.$$

If the neighbouring conductors be not earth-connected then this relation does not hold.

Suppose now that we charge an insulated conductor until it acquires unit potential. Then in the above equation V is unity, and Q is the charge required to raise the potential of the conductor to unity. But from the above equation since $V = 1$ we have

$$C = Q.$$

Hence we obtain another definition of the electric capacity of a conductor.

DEFINITION. *The* **Capacity** *of a conductor is the electrical charge required to raise the conductor—all other conductors in the field being earth-connected—to unit potential.*

We may compare this with the definition of capacity for heat[1]. The capacity for heat of a body is the quantity of heat required to raise the temperature of the body by unity; we again have an analogy between heat and electricity.

37. Energy of a Charged Conductor. If a unit positive charge be brought up to a conductor at potential V, and it be supposed that the electrical distribution is not thereby altered, the work done is by definition V; if a charge q be brought up, q being very small, so that its presence does not disturb the electrical distribution, the work done is Vq.

Now suppose we have an uncharged insulated conductor, and that we charge it by bringing up a series of equal small charges q_1, q_2, q_3, etc. Let its potential after it has received these successive charges become V_1, V_2, V_3, etc. In reality as the

---

[1] Glazebrook, *Heat*, p. 34.

charging proceeds the potential rises gradually from V_1 to V_2, V_2 to V_3 and so on. Let us however proceed to calculate the work on the assumption that the change is sudden, and that during the interval in which the charge q_2 for example is being brought up the potential is V_2 and so on. Then the work done is clearly

$$q_1 V_1 + q_2 V_2 + q_3 V_3 + \dots$$

This can be represented graphically as in Fig. 24. For let

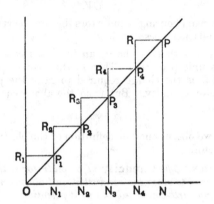

Fig. 24.

a distance ON measured along a horizontal line represent the charge Q, and let PN measured perpendicularly to ON represent the potential V. Divide ON into a large number of equal parts in N_1, N_2 etc., then these parts represent the small charges q_1, q_2 etc., and draw $N_1 P_1$, $N_2 P_2$ to represent the potentials V_1, V_2 after the respective charges have been given to the conductor. Complete the parallelograms $ON_1 P_1 R_1$, $N_1 N_2 P_2 R_2$ as shewn in Fig. 24.

Then area
$$ON_1 P_1 R_1 = V_1 q_1$$
$$N_1 N_2 P_2 R_2 = V_2 q_2 \text{ etc.}$$

Hence the work done in charging which has been shewn to be equal to

$$V_1 q_1 + V_2 q_2 + \dots$$

is equal to the sum of the areas

$$ON_1P_1R_1 + N_1N_2P_2R_2 + \ldots$$

Now if the distances ON_1, N_1N_2 etc. are sufficiently small the points P_1, P_2...etc. will lie on a continuous line $OP_1P_2...P$, and the sum of the rectangular areas $N_1N_2P_2R_2$ etc. will coincide with the area bounded by this line, and the lines ON and PN. This area then represents the work done in charging the conductor.

But since the potential is proportional to the charge it follows that $OP_1P_2...P$ is a straight line. The area then representing the work is therefore a triangle on the base ON and of altitude PN.

The area of this triangle is $\frac{1}{2}PN \cdot ON$ or $\frac{1}{2}VQ$, where Q is the sum of the small quantities q_1, q_2, etc.

Hence the work done in charging a conductor to potential V with a charge Q, all other conductors in the field being at zero potential, is $\frac{1}{2}V.Q$.

The charged conductor has acquired energy, and its energy or capacity for doing work is measured by the work done in charging it.

Hence the energy of this charged conductor is $\frac{1}{2}V.Q$.

Since we have $Q = CV$ we can express this energy E in terms either of Q or V and the capacity.

For substituting for Q we obtain

$$E = \tfrac{1}{2}VQ = \tfrac{1}{2}CV^2;$$

while substituting for V we get

$$E = \frac{1}{2}VQ = \frac{1}{2}\frac{Q^2}{C}.$$

Thus the work done in charging a conductor with a given charge is inversely proportional to its capacity, while the work done in charging it to a given potential is directly proportional to its capacity.

The method of proof should be compared with that given in Glazebrook, *Dynamics*, § 39, for finding the space passed over by a particle moving with known velocity, and the formulæ obtained with those found for the space traversed when the acceleration is constant,

viz. $s = \dfrac{1}{2}vt = \dfrac{1}{2}at^2 = \dfrac{1}{2}\dfrac{v^2}{a}.$

CHAPTER IV.

CONDENSERS.

38. Condensers. The term condenser is used for a conductor or conductors arranged to have specially large capacity.

To explain the action of a condenser let us consider the following experiment.

In Fig. 25 *A* and *B* represent two parallel vertical metal

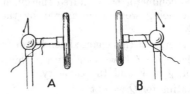

Fig. 25.

plates each insulated; the plate *A* is connected to an electroscope, the plate *B* is moveable,—its support rests in a groove, the direction of which is perpendicular to the planes of the two plates,—and it is placed at some distance from *A*. Such an instrument is called a condenser.

Electrify *A*; the gold leaves diverge, then bring *B* near to *A*, the divergence of the leaves decreases. If *B* be connected to an electroscope when at a distance from *A*, the leaves of this electroscope will diverge as *B* is brought near to *A*. Moreover it will be found that the sign of the electrification of both electroscopes is the same; this can be shewn by bringing an electrified ebonite rod near to each in turn.

Now connect B, when it is near A, to earth, the leaves of the A electroscope will collapse still further, those of the B electroscope will collapse entirely.

We can put the explanation of these observations in various ways. Thus, suppose A is electrified positively when B is brought near, negative electrification is attracted to its surface near A, and positive repelled into the gold leaves; the negative electricity attracts the positive charge of A. Part of the charge of the gold leaves connected to A is thus withdrawn, and they diverge less than originally. When B is now connected to earth, the positive charge on the back of B and on the gold leaves of the B electroscope is repelled to earth. More of the positive charge of A is, in consequence, attracted to its face, the charge on the gold leaves is reduced, and they collapse still further. It is as though the presence of the plate B, specially if it be earth-connected, condensed the positive charge of A on to the surface opposite B. Hence the name condenser. The charges on the opposing faces of the two condenser plates are sometimes spoken of as bound-charges.

We can put the matter more concisely by introducing the idea of potential.

39. Explanation of the Action of a Condenser. When the plate A is positively electrified, it and the gold leaves of the A electroscope are at a definite positive potential, the gold leaves therefore diverge; the plate B is at a lower potential, if it be earth-connected it is at potential zero. Suppose in the first case that B is insulated. As B is brought near to A its own potential rises, but the presence of B, a body at lower potential than A, reduces the potential of A, thus the gold leaves of the A electroscope become less divergent, while those of the B electroscope diverge, being raised to a positive potential by the presence of A.

If B be earth-connected its potential is zero, its presence therefore reduces the potential of A still further, and the divergence of the leaves of the A electroscope is still more reduced.

Thus in the presence of B an additional charge is required to raise the potential of A to what it was previously. The capacity of A is raised by the presence of B.

40. Capacity of a Condenser. We have already defined the capacity of a conductor as being the ratio of its charge to its potential, when all neighbouring conductors are at potential zero. When B is at some distance, the composite conductor composed of A and the electroscope has a certain capacity; on bringing the earth plate B near the total charge on this conductor—A and the electroscope—is not altered, but its potential is reduced, hence the capacity of the conductor is increased, and this increase in capacity may be very large.

If instead of using an electroscope to do the experiment with we employed a quadrant electrometer, see Section 62, or some other instrument for the measurement of potential, we might compare the potential in the two cases, (i) when B is at a distance, (ii) when it is near A, and from this comparison, if we could make sure that there had been no leak, we could determine the ratio of the capacities.

A slightly different definition of capacity from that already given, which, however, as we shall see shortly is quite consistent with it, will be found of use when dealing with a condenser. This we proceed to consider.

In a case in which all the lines of force from a body at potential V_1 pass to another body at potential V_2 it is found

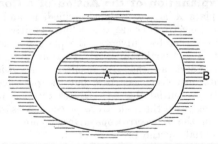

Fig. 26.

as the result both of theory and experiment that the charge on the first body is proportional to the difference of potential between the two.

The case is only fully realised when a body A (Fig. 26) containing a charge Q_1 at potential V_1 is completely surrounded by a second body B at potential V_2, when this happens it is found that the ratio $Q_1/(V_1 - V_2)$ is constant. Such an arrangement constitutes a condenser, the presence of the outer body B increases the charge requisite to raise A to a given potential. The constant ratio of the charge to the difference of potential in such a case is known as the capacity of the condenser. Hence if C be the capacity we have

$$\frac{Q_1}{V_1 - V_2} = C,$$

or
$$Q_1 = C\,(V_1 - V_2).$$

The fact that in a case of this kind we double the difference of potential between A and B by doubling Q_1 may be seen thus. Let us suppose that B completely surrounds A and that it has a charge Q_2.

Then since the charge on B produces no force at any point in the space within the inner surface of B, the force at any point in this space is entirely due to the charge on A. If we double the surface density at each point of A, and thus double its charge, we double the force everywhere between A and B. Thus the work necessary to move a unit charge from B to A is doubled. But this work is $V_1 - V_2$. Hence by doubling the charge on A we double the difference of potential between A and B, or the ratio of the charge on A to the difference of potential between A and B is a constant.

It is important to realise how the charge on B is distributed in this case. Since all the lines of force from A end on B and since the charges at the two ends of a line of force are equal and opposite, the charge on the inner surface of B must be $-Q_1$, and this charge resides exclusively on the interfaces between the conductor and the dielectric, but the total charge on B is Q_2, hence the charge on the outer surface of B must be $Q_1 + Q_2$. We may also consider the forces due to these charges. Within A the potential is constant and equal to V_1 and the force is zero. Between A and B the potential changes from V_1 to V_2, the force is entirely due to Q_1. At the inner surface of B and at all points exterior to this, the force and potential due to Q_1 on A are each equal and opposite to the force and potential due to $-Q_1$ on B. The effect of A is entirely screened by the effect due to the inner surface of B. Exterior to this inner surface, force and potential are both due to $Q_1 + Q_2$ on the outer surface of B. At points within this outer surface, points that is within the substance of B, the potential due to the charge $Q_1 + Q_2$ is constant and the force is therefore zero. At all external points the potential and the force are those which arise from $Q_1 + Q_2$ on the outer surface of B.

Now with a condenser such as we have described previously—two flat plates near together—the above conditions
do not hold; all the lines of force from A do not pass to B.
But the condition is very nearly realised when the plates
are near together, the number of lines of force from the back
of the plate A is very small, almost all issue from its front
surface and fall on B.

Hence in this case also we may say that the ratio of the
charge to the difference of potential is constant, and define
this ratio as the capacity of the condenser thus:

DEFINITION. *The* **Capacity** *of a condenser is the ratio of
its charge to the difference of potential between its plates.*

If we call the capacity C we have
$$Q = C\,(V - V'),$$
where Q is the charge, V and V' the potentials.

We may notice that if V' be zero so that all neighbouring
conductors are at zero potential we have $Q = CV$, the same
equation as previously, the two definitions are consistent.

41. Energy of a Charged Condenser. We know
that the energy of a charged conductor is $\frac{1}{2}QV$. Hence in
the case of the condenser in which we have one plate with
charge Q and potential V and another with charge $-Q$ and
potential V' the energy is
$$\tfrac{1}{2}QV - \tfrac{1}{2}QV'.$$
This may be written $\frac{1}{2}Q\,(V - V')$ or $\frac{1}{2}C\,(V - V')^2$,

or
$$\frac{1}{2}\frac{Q^2}{C}.$$

42. Inductive Capacity. EXPERIMENT 15. *To shew
that the capacity of a condenser depends on the nature of the
dielectric between its plates.*

Arrange the condenser AB (Fig. 25) so that the plate A is
connected to an electroscope while B is earthed. Take a plate
of glass or other insulating material, making sure that its
surfaces are unelectrified by passing it over a flame, and place
it between the plates of the condenser, taking care not to

touch the insulated plate[1], the divergence of the gold leaves is reduced; a change of the dielectric has increased the capacity of the condenser, and reduced the potential of the insulated plate.

DEFINITION. *The ratio of the capacity of a condenser having a given material for its dielectric[2] to the capacity of the same condenser with air for the dielectric is called the* Specific Inductive Capacity of the Condenser.

The term specific inductive capacity has been already defined, see Section 27, and we have learnt that the force between two charges e, e' at a distance r apart in a medium of inductive capacity K is ee'/Kr^2. It follows from theory and can be shewn directly by experiment that the two definitions are consistent. The quantity K given by the above equation is the inductive capacity as defined by the definition of this Section.

43. Calculation of Capacity. The capacity of a condenser does not depend on its charge or its potential, but merely on its size and shape and the distance apart of its surfaces. We can calculate it for the plate condenser thus if we assume a relation which is readily proved as a consequence of the inverse square law, but of which the proof lies rather outside our limits.

The relation is that, if in a medium of inductive capacity K, R be the resultant force just outside a conducting surface, on which the surface density is σ, then $K \cdot R = 4\pi\sigma$, or in words, the product of the resultant force and the inductive capacity is equal to 4π times the surface density. This is known as Coulomb's law.

Now in the plate condenser if the distance apart of the plates is small compared with their size the field of force is uniform and the force is constant. Hence if c be the

---

[1] Theoretically contact with the insulated plate ought not to matter, it is difficult however to prevent some leakage over the surface of the dielectric, and this if it occurred would simulate the result sought for.

[2] In this definition the dielectric is supposed to replace the air completely.

distance apart of the plates, V, V' their potentials, the work done in carrying a unit charge across is Rc but it is also

$$V - V'.$$

Hence
$$Rc = V - V',$$

or
$$R = \frac{V - V'}{c}.$$

But if Q is the charge, S the area of the positive plate and C the capacity, then

$$Q = S\sigma,$$

also by Coulomb's law
$$KR = 4\pi\sigma.$$

Hence
$$R = \frac{4\pi\sigma}{K} = \frac{4\pi Q}{KS}.$$

Therefore
$$\frac{4\pi Q}{KS} = \frac{V - V'}{c}.$$

Hence
$$C = \frac{Q}{V - V'} = \frac{KS}{4\pi c}.$$

Thus if we know the area and distance apart of the plates and the inductive capacity of the dielectric we can calculate the capacity of the condenser.

If C_0 be the capacity of the same condenser with air as dielectric, since for air $K = 1$, we find

$$C_0 = \frac{S}{4\pi c}.$$

Hence from these two results we obtain

$$K = \frac{C}{C_0},$$

and this agrees with the definition of specific inductive capacity given in Section 27 above.

44. Leyden Jar. Condensers take various forms. One of the most usual is that of the Leyden Jar. This as usually constructed is shewn in Fig. 27. A wide-mouthed glass bottle is coated in part within and without with tinfoil. The glass above the tinfoil is coated with shellac varnish to maintain the insulation; the bottle is closed by a wooden stopper through which runs a piece of brass rod; the upper end of this carries a knob or ball, the lower end is in connexion by means of a light chain or a piece of wire with the inner coating of tinfoil. The two coatings of tinfoil constitute the plates of the condenser; the glass is the dielectric, the inner coating is insulated by means of the glass, the outer coating can be insulated if necessary by placing the whole instrument on an insulating stand.

Fig. 27.

To charge the jar the knob is brought near to an electrical machine or other source of electricity, sparks pass across from the machine to the jar and a considerable quantity of electricity can be communicated to the inner coating.

If a connexion be made between the two coatings by means of the discharging tongs shewn in Fig. 28, one ball of which is made to touch the outer coating while the other is brought near to the knob, a spark passes. To secure the observer from shock the handle of the tongs is of glass or some other insulator.

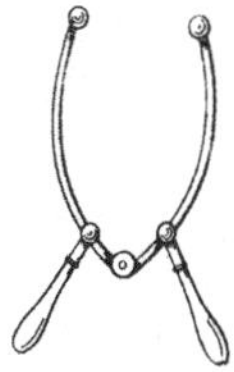

The hygroscopic qualities of various glasses differ greatly; it is important to choose for a Leyden Jar a glass which is not hygroscopic.

Fig. 28.

Another form of condenser, Fig. 29, consists of a number of sheets of tinfoil insulated from each other by alternate

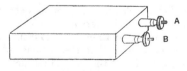

Fig. 29.

sheets of mica or of paraffin paper. The odd sheets of the tinfoil, the first, third, fifth, etc. are connected together and form one plate of the condenser, the even sheets, the second, fourth, sixth, etc. are also connected and form the other; the whole is mounted in a case, the two plates being connected to the terminals A and B.

45. Batteries of Leyden Jars. The capacity of a condenser is, we have seen, proportional to the surface of one of its plates and inversely proportional to the distance between them; it can be increased then either by reducing this distance or by increasing the area of the surface. There are limits however below which we cannot reduce the distance, for if the dielectric be too thin it is pierced by an electric spark when the potential difference is raised. We can however practically increase the surface by connecting up a number of jars as shewn in Fig. 30. The outer plates are all

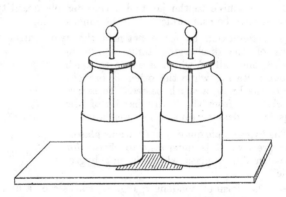

Fig. 30.

connected together by placing the jars on a sheet of tinfoil or metal; the knobs of all the jars are in electrical communication and thus the inner coatings are also connected. In such an arrangement it is clear that the capacity of the whole is the sum of the capacities of the individual jars.

For some purposes jars are connected as shewn in Fig. 31, they are then said to be "in cascade." The outer coating of the first jar is connected to the inner coating of the second and so on in succession. All the jars except the last have their outer as well as their inner coatings insulated, the outer coating of the last jar is to earth.

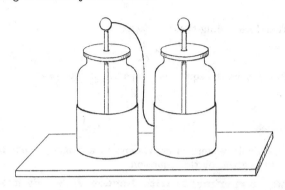

Fig. 31.

Suppose a charge Q given to the inner coating of the first jar, the one to the right hand in the figure. Let its potential be V_1 and the potentials of the successive inner coatings be $V_2, V_3 \ldots V_n$. The potential of the outer coating of the nth or last jar is zero.

The charge Q induces $-Q$ on the outer coating of the first jar and repels Q to the inner coating of the second, and this is continued throughout the system. Thus the charges of all the jars are the same. Let $C_1, C_2 \ldots C_n$ be their capacities.

Then we have

$$Q = C_1 (V_1 - V_2),$$
$$Q = C_2 (V_2 - V_3),$$
$$\cdots\cdots\cdots\cdots\cdots$$
$$Q = C_n (V_n - 0).$$

Whence

$$V_1 - V_2 = \frac{Q}{C_1},$$

$$V_2 - V_3 = \frac{Q}{C_2},$$

$$\cdots\cdots\cdots\cdots$$

$$V_{n-1} - V_n = \frac{Q}{C_{n-1}},$$

$$V_n - 0 = \frac{Q}{C_n}.$$

Therefore adding these all together

$$V_1 = Q\left\{\frac{1}{C_1} + \frac{1}{C_2} + \dots + \frac{1}{C_n}\right\}.$$

But if C be the equivalent capacity of the system

$$V_1 = \frac{Q}{C}.$$

Hence

$$\frac{1}{C} = \frac{1}{C_1} + \frac{1}{C_2} + \dots + \frac{1}{C_n}.$$

The whole system is equivalent to a single jar having a capacity C given by this equation.

46. Experiments with Leyden Jars. By making the coatings of a Leyden Jar removable it can be used to shew that the charge of a conductor resides on the surface of the dielectric which insulates it. This was first done by Benjamin Franklin. The jar as shewn in Fig. 32 is in the form of a tumbler; the coatings are both of tin or brass; the inner coating can be lifted out of the glass and then the glass can be removed from the outer coating. The jar is charged in the usual way, on removing the coatings and examining them they are found to be uncharged; the glass when examined by aid of an electroscope is found to be strongly charged.

47. The Condensing Electroscope. In this instrument, Fig. 33, the action of a condenser is applied to render sensible the action on a gold-leaf or other electroscope of a source of electricity at low potential. Such a source, if connected to the electroscope directly, is not sufficient to produce any visible effect on the leaves.

Fig. 32.

Instead of a knob the electroscope is fitted with a flat plate, the upper side of which is varnished so as to make it insulate. An insulating handle is attached to a second similar plate which is varnished on the lower side. On placing this plate on the electroscope the two constitute a condenser of large capacity, the dielectric being the thin layer of varnish.

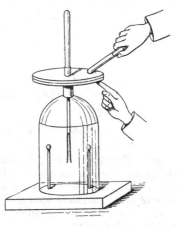

Fig. 33.

The upper plate is now connected to the electrified body and the lower plate is earthed for a moment. The difference of potential between the plates is not large, but in consequence of their large capacity a considerable charge, of the opposite sign to that on the body, is communicated to the lower plate and remains there when the earth contact is broken; the lower plate and the electroscope are then at potential zero. Now remove the upper plate, which for the moment we will assume to be positive, in consequence of the removal of the positive plate the potential of the lower plate with its negative charge falls considerably below that of the cage or tinfoil strips on the glass cover; the leaves therefore diverge with negative electricity. The small difference of potential between the plates which exists when the condenser has a large capacity is increased many times when the upper plate is removed and the condenser capacity in consequence reduced.

By means of this condenser action the gold-leaf electroscope may be used to indicate very small differences of potential.

CHAPTER V.

ELECTRICAL MACHINES.

48. Frictional Machines. The only means of obtaining an electric charge described up to the present has been by the friction of two dielectric materials. Various forms of apparatus have been devised with a view of obtaining larger charges than it is possible to do by the friction of a rod of glass or of ebonite.

49. The Plate Electrical Machine. This, which is shewn in Fig. 34, consists of a circular plate of glass mounted so as to rotate about a horizontal axis. Two pairs of cushions rub against the glass at opposite extremities of a vertical diameter, and two U-shaped insulated conductors with points on the side towards the glass are placed to collect the electricity produced by the friction. The points are known as the combs.

These two conductors are connected together and also to an insulated conductor known as the prime conductor of the machine.

Two opposite quadrants of the plate between the cushions and the combs are usually covered with flaps of oiled silk. The cushions are of wash-leather stuffed with horsehair, covered with an amalgam of tin and zinc with mercury smeared on to them with a little lard or tallow. To secure good working the cushions should be earth-connected. The connexion afforded by the wood framing of the machine is usually sufficient; in some cases however contact is obtained by a piece of chain or by strips of tinfoil. The plate is turned by a handle, the direction of motion past the cushions being from the uncovered towards the covered quadrants.

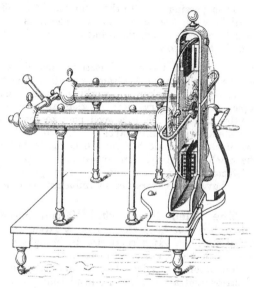

Fig. 34.

When the machine is in action the friction of the cushions produces positive electrification on the glass, negative on the amalgam; the negative escapes to the earth, the positive is carried forward on the glass towards the combs. As this positive electricity approaches the combs the prime conductor becomes electrified by induction; negative electrification is attracted to the combs and positive repelled to the other end of the conductor. The electrical force at the points becomes very great and a wind of negatively electrified particles of air blows from them on to the glass; the glass plate is thus discharged; its electrification has passed to the prime conductor; the unelectrified glass becomes charged again as it passes through the cushions and the process is repeated.

The silk flaps attached to the cushions become negatively electrified, and by their action tend to produce a more uniform slope of the potential along the surface of the glass than would otherwise be possible ; the charge on the glass is thus prevented from passing back to the rubber and the difference of potential between the rubber and the combs is increased. Sparks can be collected from the prime conductor by bringing another conductor up to it.

There are various forms of electrical machines. The principle however is much the same for all. Various precautions have to be taken in working them. Glass is hygroscopic ; it is necessary therefore that all glass surfaces should be dry and clean ; it is a good plan to coat the glass insulating stems with a thin layer of varnish made by dissolving shellac in pure alcohol ; before use it is desirable to warm the machine slightly. In consequence chiefly of these various defects frictional machines are now practically obsolete ; they have been superseded by influence machines.

50. The Electric Spark. Various experiments already described can be performed on a larger scale with an electrical machine. If a conductor be brought near to a machine which is being worked, the air between the machine and the conductor is subjected to a gradually increasing electrical stress, and after a time its insulating power is overcome and it gives way, a spark passes from the machine to the conductor. The light from the spark is due to the very great rise in temperature produced by its passage, the air is hereby rendered incandescent. As the machine is worked sparks continue to pass until the potential of the conductor has risen greatly and has approached so nearly to that of the machine that the difference between the two is insufficient to rupture the air.

When the conductor of the machine has been raised to a high potential and the machine continues to be worked, the electricity is discharged into the air ; the air particles near the machine are charged and the forces on them become very great ; the particles are repelled from the machine to have their places taken by others and an electric wind is set up.

In the dark the positive conductor of the machine is surrounded with a violet glow which is specially marked near
any angle or sharp point. To prevent this loss all points or
angles should be avoided as far as possible. Any points on
the negative conductor shew stars of white light.

The effect of points in discharging electricity may be
illustrated in various ways. Thus with a given machine in
its normal state it may be possible to draw sparks an inch or
more in length; if a fine point be affixed to the conductor the
length of spark is reduced to a small fraction of what it was
previously and the violet glow of the positive discharge is
seen from the point. If a lighted candle be held as in
Fig. 35 near the point, the flame is violently blown away.

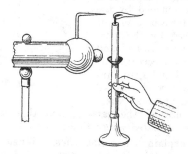

Fig. 35.

The electric windmill is another example of the action of
points. A number of pins with their points turned in the
same direction are attached like the spokes of a wheel to a
small central cup and balanced on a pivot connected with the
conductor of an electrical machine; on working the machine
the pins revolve, moving round in the opposite direction to
that in which the points are bent. The electric wind which
blows from each point reacts on the mill and drives it round.
If the points be blunted by sticking on to each a bit of sealing-
wax the mill will no longer turn.

More or less successful attempts have been made to utilize
this electric wind on a large scale to remove metallic fumes or

smoke from the air. If the air near a discharging point is loaded with small particles, these are driven away and deposited on opposing surfaces.

51. The Electrophorus. The instrument, Fig. 36, consists of a flat smooth plate of resin, ebonite or some other insulating substance—the cake —which rests on a metal plate —the sole—and of a second metal plate—the cover—which is supported by an insulating handle. To use the instrument the cake is electrified by friction with catskin, receiving hereby a negative charge; the cover is laid on it and touched by the experimenter for a moment; it is then removed by the insulating handle and will be found to be positively electrified; if it be discharged and again laid on the cake a second positive charge can be obtained with-

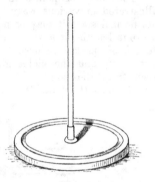

Fig. 36.

out renewed use of the catskin, and this can in a dry atmosphere be repeated many times.

We may explain the action thus. When the cover is placed on the electrified cake it rests on a few points only, elsewhere the two are separated by a thin layer of air. The electrification of the cake acts inductively on the cover. When the cover is touched positive electricity passes on to it from the finger, the outer field is destroyed and the cover is left with a positive charge. The distribution is as shewn in Fig. 37. On removing the cover this positive charge is carried away with it, the cake is left, except for leakage, with practically the same charge as it possessed previously, and the cover

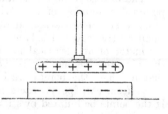

Fig. 37.

after discharge can be replaced and the operations repeated.

The action of the sole tends to reduce the leakage into
the air and thus to prolong the period during which the
effect of a single application of the catskin is effective. For
the sole being in contact with the earth is necessarily at zero
potential, the strongly electrified ebonite would if the sole
were not present be at a considerable negative potential, and
the force tending to discharge it would be large; the presence
of the sole reduces the difference of potential between the
cake and the air near it and thus reduces the tendency
to leak.

It is interesting to trace the distribution of the lines
of force during these various processes. When the cake
receives its negative charge the sole becomes positive. Lines
of force pass from it through the ebonite to the upper surface
of the cake; the field is almost entirely confined to the ebonite,
though a few lines may pass from the table and walls of the

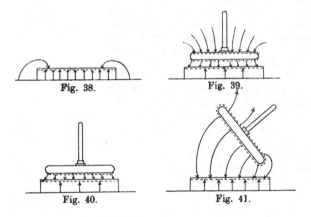

Fig. 38.   Fig. 39.

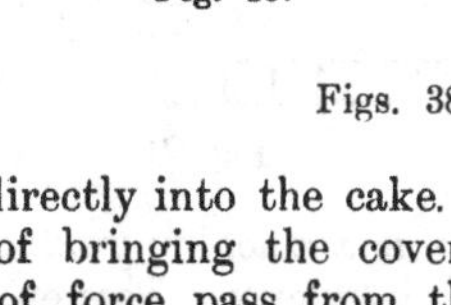

Fig. 40. 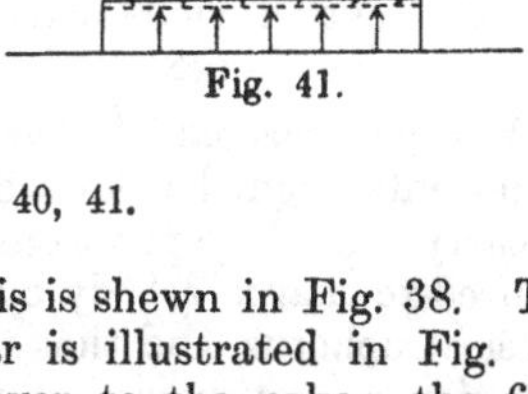   Fig. 41.

Figs. 38, 39, 40, 41.

room directly into the cake. This is shewn in Fig. 38. The
effect of bringing the cover near is illustrated in Fig. 39.
Lines of force pass from the cover to the cake; the field

is transferred in great measure from the ebonite to the air-space between the cake and the cover, but the upper surface of the cover being negatively charged receives from the walls and surrounding objects lines of force equal in number to those which pass from it to the cake; the number of lines traversing the ebonite is small and the positive charge of the sole is now distributed over the walls, there is an external field due to this charge and the negative charge on the upper side of the cover. When the cover is touched this external field is destroyed, the field in the air-gap between the cake and the cover remaining much as before, and we have the condition shewn in Fig. 40.

As the cover is removed the lines between it and the cake lengthen, some of them bulge outwards as shewn in Fig. 41, and finally come into contact with the walls; here they break into two parts, one of which shortens again until it passes directly through the ebonite between the sole and the cake, while the other ends in a negative charge on the walls. This is shewn in Fig. 42. As the cover is moved further away this happens to an increasing number of the lines between it and the cake, until finally they are practically all broken and we are left, so far as the cake is concerned, with the same state of affairs as after the application of the catskin, but in addition there is now another field due to

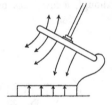

Fig. 42.

the charge on the cover. This is shewn in Fig. 43. Electric energy is stored in this field, and this energy is derived from the additional work which has been required to lift the cover from the cake in consequence of its positive charge.

Various mechanical arrangements might be devised to carry out the operations required to charge a body by

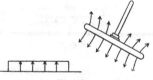

Fig. 43.

the electrophorus, and thus to give us a machine producing by induction a continuous supply of electricity. This end however is attained more easily by means of some of the induction machines we are about to describe.

52. Influence Machines. The principle of all these is much the same. In their simplest form there are two fixed insulated conductors A and B which we may call the collectors, and a moving conductor C, which can, after being electrified, be brought near to A and B in turn in such a position that if contact be established or a spark allowed to pass, electricity will be transferred from C to A or B as the case may be, and C will move on unelectrified.

Suppose now that we start with a slight difference of potential between A and B, A being say slightly positive and B negative. Bring C near to A and connect it to earth; it is electrified negatively by induction. Break the earth connexion and move C on until it is near to B. Connect B and C, the negative charge on C passes to B, which becomes more negative. Break this connexion, and while C is still near to B connect it to earth ; it receives a positive charge by induction. Break this earth connexion and bring C near to A, putting the two into connexion; the positive charge passes to A which becomes more positive. We have thus come to the end of a cycle ; on breaking the connexion between A and C and again putting C to earth, it receives for a second time a negative charge and the whole series can be repeated. But we must notice that this second negative charge is greater than the first, for by the action of the machine A has received an increase in its positive charge and therefore induces a larger charge on C when near it than it did at first. Thus not only is the potential difference between A and B continually increased by the action of the machine, but also the rate at which it is increased grows rapidly.

Nicholson's revolving doubler, invented in 1788, was probably the first machine of this kind; since that time it has taken various forms, and of these we will describe one or two.

53. The Replenisher. In this instrument (Figs. 44 and 45), designed by Sir William Thomson (Lord Kelvin) in 1867, the two collectors A and B are two portions of a cylinder, the axis of which cuts the paper at right angles at O. Each of

these is insulated, and from the inside of each there projects
a small spring, a, b. There are two carriers C_1, C_2, which are
portions of a second smaller cylinder. These are connected

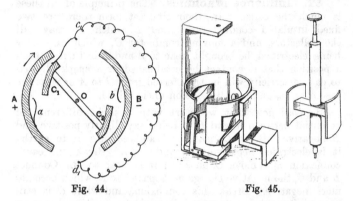

Fig. 44.   Fig. 45.

together by an insulating bar carried by an ebonite rod which
can be made to turn about the axis of the outer cylinder.
Two springs c_1, d_1 connected together by a wire which need
not be insulated, touch the carriers when these are in a
position to be affected by the inductive action of charges on
A and B. As the carriers turn still further in the direction
of the arrow, contact with c_1, d_1 is broken, and contact with b,
a is made.

Suppose now that there is a small initial difference of
potential between A and B, produced if need be by electrifying A by induction from an ebonite rod so that A is positive
with respect to B.

Then negative electricity is induced on the carrier C_1, and
positive is repelled to C_2. The carriers are so placed that
a motion in the direction of the arrow from the position
shewn breaks the contacts with c_1 and d_1, the positive
and negative charges then are insulated from each other, the
negative charge on C_1 is carried round towards B, the positive
on C_2 towards A. When C_1 is in contact with b it is approximately surrounded by the conductor B and a large fraction
of its negative charge passes to the outside of that conductor;

the converse happens to C_2. As the motion continues C_1 comes into contact with d_1 and C_2 with c_1; in consequence of the inductive action of A and B, the carrier C_1 becomes positive and C_2 negative, and the process is repeated. The difference of potential between A and B is thus continually increased and that at an increasing rate. A conductor connected to A becomes increasingly positive, one connected to B increasingly negative, and this continues until the insulation gives way somewhere and a spark passes. The instrument in this form, however, is not used to produce large charges of electricity, but rather to maintain some portion of another instrument at a given potential. As the potential falls by leakage it is raised by giving the replenisher a few turns.

Fig. 46.

54. Wimshurst's Machine. As an example of a powerful influence machine we will describe Wimshurst's. This is shewn in Fig. 46.

It consists of two plates of glass which carry narrow strips of tin-foil arranged radially at equal distances apart. The plates, which are coated with shellac to preserve their insulating power, rotate in opposite directions about a horizontal axis. At the opposite ends of the horizontal diameter are two U-shaped conductors furnished with points on the sides towards the plates; each of these is insulated and connected with a Leyden jar or other condenser. (See Section 44.)

Discharging rods fitted with insulated handles are attached to the conductors. Two diametral conductors which need not be insulated are placed as shewn in the figure, being approximately at right angles to each other. These carry brushes, which just touch the tin-foil strips as they pass under them.

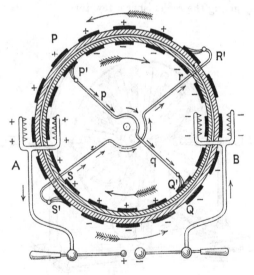

Fig. 47.

The action of the machine is best explained by the diagram Fig. 47, in which, following a suggestion due to Prof. S. P. Thompson, the rotating plates are represented as though they were two cylinders of glass rotating in opposite directions. The tin-foil sectors are shewn as dark lines on the surface of the cylinders, the diametral conductors *pq*, *rs* occupy the position indicated, and the cylinders rotate in the directions of the arrows; the inner cylinder, which corresponds to the front plate of Figure 46, having a right-handed, and the outer cylinder a left-handed motion. Now suppose that one of the strips, *P*, on the upper side of the outer cylinder becomes positively electrified; as it rotates it will be brought opposite to a strip *P'* on the inner cylinder at the moment when this is in contact with the brush *p*. This strip then acquires a negative charge, the strip *Q'* in contact with *q* being simultaneously charged positively; as the rotation continues the induced negative charge on *P'* is carried towards the right-hand comb *B*, the positive charge on *Q'* towards the left-hand comb *A*. Consider, however, the original strip *P*, in its motion it is brought under the left-hand comb *A*. A negative wind blows from the comb discharging *P*, and *A* acquires a positive charge while *P* moves on uncharged. Meantime *P'* with its negative charge and *Q'* with its positive have been brought opposite to strips at *R'*, *S'*, which are connected together by the second diametral conductor *rs*. *R'* hereby acquires a negative charge and *S'* a positive one which are carried on towards *B* and *A* respectively, while *P'* and *Q'* continue their motions. As *P'* comes under the comb *B* a positive wind is produced from the points and *B* receives a negative charge, *A* at the same time receiving a positive charge from *Q'*. *R'*, *S'* again in their motion have been brought opposite to the diametral conductor *pq*, and after inducing charges on the strips then in contact with this conductor have passed on to deliver their charges to *B* and *A*. In this way the difference of potential between *B* and *A* is continually increased.

It will be noticed that the upper part of the outer cylinder is always carrying positive electricity from *r* to *A*, while its lower part carries negative from *s* to *B*. Similarly the upper part of the inner cylinder carries negative from *p* to *B*, the lower part carries positive from *q* to *A*.

In working with a Wimshurst machine it is not usually necessary to charge one of the strips to start with, the slight friction between the brushes and the strips, or the small residual charge retained from previous use, is generally sufficient to start the machine.

The object of the Leyden jars is to increase the quantity of electricity given by each spark ; the sparks pass, for a given position of the knobs, when the potential difference between them exceeds a certain limiting value. Now the Leyden jars increase the capacity of the conductors, they increase therefore the quantity which is required to charge them to a given difference of potential. Each strip as it passes the combs conveys a definite charge, and it will need the passage of a larger number of strips to produce a spark if the jars are on than is required when they are off. Thus the sparks occur less frequently with the jars on than without them. But at each spark the conductors are completely discharged and the charges are increased by connecting up the jars ; each spark contains a larger quantity of electricity than would pass if the jars were removed.

In large machines several pairs of plates are mounted together on the same axis.

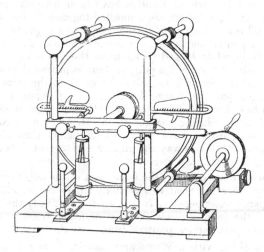

Fig. 48.

55. Holtz Machine. The Holtz Machine is older than the Wimshurst, and in consequence of its liability to fail in action in damp weather is less used at present; a brief description therefore will suffice.

The machine is shewn in Fig. 48. It consists of two glass plates, one fixed, the other capable of rotation about a horizontal axis. At opposite extremities of a diameter of the fixed plate two holes or "windows" are cut in the glass, and two pieces of varnished cardboard—a bad conductor—are fastened to the glass on the side remote from the moving plate. Each of these cardboard armatures is provided with a point or tongue, which projects through the window, so as almost to graze the moving plate; the one tongue points upwards, the other downwards. Opposite to these points, but separated from them by the moving plate, are two combs connected to the conductors of the machine. These can be placed in contact by means of the brass rod fitted with an insulating handle shewn in the figure. The machine is not self starting; to excite it a charged piece of ebonite is held near one of the armatures, and the conductors brought into contact. The moving plate is then set into rotation, the direction of motion being towards the points, and after a little time a peculiar hissing sound is heard; while the combs become slightly luminous. The ebonite may now be withdrawn, and on separating the two conductors sparks will pass between them.

The action of the machine can best be explained by the diagram (Fig. 49) in which the moving plate only is represented. P and Q are the armatures, A and B the conductors. When the negatively charged ebonite E is brought near, the conductors being connected as at a, b, positive electricity is induced on the positive comb A, and negative is repelled to B. In consequence of the action of the points these two electrifications are discharged on to the glass, and the moving plate becomes positively electrified near A, negatively electrified near B, and this continues as the plate is turned until the whole of its upper half is positively and the whole of its lower half negatively electrified. As the motion continues the positively charged portion of the plate comes close to the tongue Q, while the negatively charged part approaches P. The armature Q therefore discharges negative electricity on to the back of the plate,

receiving itself a positive charge, while P becomes negative.
These charges on the armatures now act on the combs in
the same manner as the original charge on the ebonite did,

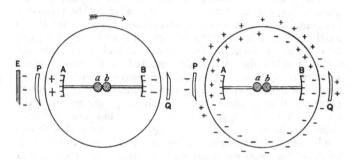

Fig. 49.

and the action continues; moreover since the charges on P
and Q are continually increasing the potential difference
between the combs rises rapidly, and if the knobs a, b are
separated it may be sufficient to produce a spark between
them, and so permit of the transference of positive electrification from A to B which is necessary for the action of the
machine. Leyden jars are sometimes attached to the machine,
and they act as in the Wimshurst. Sparks pass less frequently, but the quantity of electricity conveyed in each
spark is increased. If the knobs are separated so far that
the sparks cease to pass, the necessary transfer from A to B
cannot take place, negative electricity is carried round
on the glass to the positive armature, while positive reaches
the negative armature, the charges on the armatures are
reduced, they may even change sign, and the machine may
cease to work; if however the knobs are put together again,
so that the transfer can go on, the usual action is resumed. It
may however happen that by this the poles are reversed, for if
the knobs have been separated too long the negative armature
may have become positive and the positive negative.

Various modifications of the machine have been introduced
with a view to prevent this action, but into these it is not
necessary to enter.

56. The Voss Machine. This machine acts in the same manner as the Replenisher shewn in Figure 44. The collectors A, B are replaced by two pieces of tin-foil cemented to the back of a fixed disc, the carrier is represented by a number of studs on the opposite sides of a movable disc. To the springs a, b correspond two brushes secured to the collectors, and arranged to make contact with the carriers as they pass round. The springs $c_1 d_1$ are represented by a diametral conductor, with brushes of thin brass wire which connect together two opposite studs when they are under the influence of the collectors. The electricity produced passes from the collectors through the points on the combs to the conductors of the machine. The action is the same as in the case of the Replenisher, the friction between the studs and the brushes is usually sufficient to produce some slight potential difference between the collectors, suppose A is positive. A stud C_1 when influenced by A is connected through the diametral conductor with an opposite stud C_2. Thus C_1 becomes negative, C_2 positive; as the disc rotates C_1 comes in contact with b, and part of its negative charge passes to B; when it again reaches the diametral conductor C_2, it is under the inductive influence of B and becomes positive, and this positive charge is communicated in part to A through the brush a. Thus the difference between A and B is continually increased.

57. The Water-dropping Accumulator. As another example of an influence machine Lord Kelvin's water-dropping accumulator (Fig. 50) may be mentioned.

It consists of an insulated hollow cylinder A, which we will suppose to be charged positively. A metal pipe B connected to earth, having a fine nozzle, projects down the axis of this cylinder to about its centre; below is a second insulated hollow cylinder C, also of metal, containing a funnel. Water is allowed to drop slowly from the nozzle into the funnel, from which it escapes. Each drop on leaving the nozzle is negatively electrified owing to the inductive influence of A. The drop carries with it to the funnel its negative electrification: when in contact with the funnel it is practically in the interior of a hollow closed conductor, it gives up its charge to the cylinder C

and escapes unelectrified. This process could go on continuously, except for the fact that the moisture of the air would soon destroy the electrification of A, but by putting up the apparatus

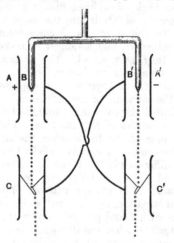

Fig. 50.

in duplicate, and connecting A to C', A' to C, if we start as before with A positive, A' will receive a negative charge from C. This will act inductively on the water dropping through it, thus C' will become positive, some of the positive charge received by C' will pass to A, and as a result the difference between A and A' will go on increasing.

CHAPTER VI.

MEASUREMENT OF POTENTIAL AND ELECTRIC FORCE.

58. Electroscopes and Electrometers. We have described various simple forms of apparatus for indicating the presence of electricity, such as the pith ball, or the gold-leaf electroscope; we must now consider some more delicate appliances which can be used for measurement, and not merely as indicators of potential difference.

59. Coulomb's Torsion Balance. The earliest such instrument was Coulomb's torsion balance. This, which is shewn in Fig. 51, consists of a small gilded pith ball A, at the end of a long light rod or finger ABC of some insulating material; Coulomb used shellac. This rod is suspended within a glass case in a horizontal position by means of a very fine vertical wire BD. The wire passes up a vertical glass tube attached to the case of the instrument, and is fixed at D to a torsion head. The tube carries a horizontal circular plate at its upper end, the edge of this plate is graduated, and the torsion head, which can be turned about a vertical axis through the centre of the plate, has a pointer attached so that its position can be read off on the circle. The case of the instrument is usually circular, and the position of the balance ball can be observed by the aid of a graduated circle engraved on the case. A second small ball E carried on an insulated stem can be introduced through a hole in the case and brought into any desired position near A.

If this ball E be charged and introduced near A attraction at first takes place; A then receives a portion of the charge on E and is repelled; this produces a twist in the wire, which

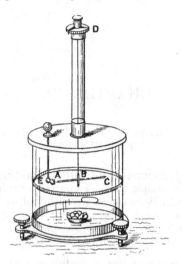

Fig. 51.

is resisted by its elasticity, and A comes to rest in a position in which the electrical force of repulsion is equal to the force arising from the twist of the wire. But this latter force is known to be proportional to the angle through which the wire is twisted, and this latter angle is given in terms of the graduations on the case by observing the original and final positions of A; if then we find that in one experiment the angle of twist is $a°$ and in another $\beta°$, we know that the forces are in the ratio of a to β. It was with an instrument of this kind that Coulomb proved the law of the inverse squares, and made very many other important measurements. Not the least interesting among the historical exhibits in the Paris Exhibition of 1900 was Coulomb's original torsion balance.

The torsion balance has now been superseded by more

delicate instruments, but the general idea of its use to verify the law of inverse squares may be given thus. Suppose that in the above experiment the shellac carrier supporting the ball A is twisted through an angle $a°$; the distance between the balls A and E, if E be placed in the position originally occupied by A, will be approximately[1] proportional to a.

Now by twisting the torsion head in the opposite direction to that in which A has moved, the distance between the balls can be reduced. Let us reduce the angular distance to $\frac{1}{2}a$, half its previous value, and suppose that in order to do this we have to turn the torsion head through an angle β. The total twist on the wire is then $\beta + \frac{1}{2}a$, for the upper end has been twisted through an angle β, while the lower end was turned in the opposite direction through an angle a and then brought back through $\frac{1}{2}a$.

Thus the twist is $\beta + a - \frac{1}{2}a$, or $\beta + \frac{1}{2}a$.

Now the forces on the balls are proportional to the angles of twist.

Hence[2]

$$\frac{\text{Force in position 2}}{\text{Force in position 1}} = \frac{\beta + \frac{1}{2}a}{a}.$$

[1] If a be the length AB, the distance apart of the two balls is $2a \sin \frac{1}{2}a$, and if a is not large this is approximately proportional to a. In careful work however this formula should be used.

[2] The formulæ may be put more accurately thus.

Let a be the angular distance between the balls in the first instance, γ in the second, and β the angle the upper end of the wire has been turned through; the angles of twist of the wire to which the forces are proportional are a and $\beta + \gamma$ respectively; the distances between the balls are $2a \sin \frac{1}{2}a$ and $2a \sin \frac{1}{2}\gamma$.

Thus it is found by making a number of observations that

$$a \sin^2 \frac{1}{2}a = (\beta + \gamma) \sin^2 \frac{1}{2}\gamma.$$

Thus if we call F_1, F_2 etc. the forces in the various positions, r_1, r_2 the distances apart, we see that

$$F_1 r_1{}^2 = F_2 r_2{}^2 = \ldots\ldots,$$

or if F be the force at distance r the product Fr^2 is constant for all distances.

Hence it follows that F is inversely proportional to r^2.

But on making the experiment it is found that approximately $\beta = 3\frac{1}{2}a$.

Thus $\beta + \frac{1}{2}a = 4a$, and the ratio of the forces is four to one, or by halving the distance between the balls the force is quadrupled. If the distance is reduced to one-third of its original value the force is found to be increased nine times, and so on.

Hence the force is inversely proportional to the square of the distance.

As an example we might note that the angle between the balls in the first position was 36°. On twisting the upper end of the wire until the angular distance was halved, we should find that the twist required was about 126°. The total twist on the wire would then be 126° + 18° or 144°; but this is four times 36°. Hence we infer that the force in the second position is four times that in the first.

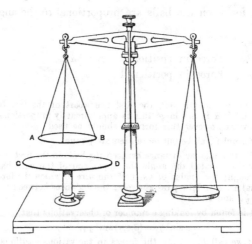

Fig. 52.

60. The Attracted Disc Electrometer. Let AB, Fig. 52, represent a circular disc supported horizontally from one arm of a balance and counterpoised if necessary by weights in the other pan, and let CD be a second disc fixed in a horizontal position at a small distance from AB and insulated; let AB be connected to earth. If now CD be electrified AB will be attracted and weights must be put into the opposite pan to maintain the beam in its horizontal position. There will be a relation between these weights, which measure the attraction between the two discs, and the potential to which CD has been electrified, and by observing the weights the potential of CD can be calculated from this relation.

This is the principle of the attracted disc electrometer.

If we knew the distribution of the lines of force between AB and CD it would be possible to calculate the attraction on AB. With two discs as described it would be difficult to determine this distribution; the lines of force in the centre of the discs would run in straight lines from one disc to the other; at the edges however they would bulge outwards and the calculation of the attraction would be troublesome if not impossible. This difficulty is avoided by making the moveable disc AB the central portion of a very much larger plate; the disc is separated from the plate by a very narrow aperture just sufficiently wide to allow the disc to move freely through it. In this case we may treat the lines of force over AB as straight lines perpendicular to AB. The fixed outer portion of the disc AB is known as the guard-ring.

In the above we have supposed the measurement made by determining the weights required to balance the electrical attraction, we may use the instrument in a different manner; clearly if we decrease the distance between the two discs we increase the attraction and *vice versâ*. Now let us suppose that before CD is brought near, the weights in the scale-pan are slightly too heavy, so that the pointer of the balance is a little to the left of the centre of the scale. Then on electrifying CD and bringing it near, we attract AB and we can determine the position of CD required to bring the pointer to the centre of the scale, into the sighted position we may call it.

Let us suppose that the potential of CD is V and the distance between the discs is a. Suppose now that CD is electrified to a different potential V' and that when AB is again brought into the sighted position the distance apart is a'. We know that the attraction on the disc AB is the same in these two positions, for the weight has not been changed; but this attraction depends on the resultant electrical force acting at each point of AB.

Let us draw the equipotential surfaces in the two cases, in each case they will be planes parallel to the disc, and in each case the consecutive surfaces will be at a constant distance apart; in the first case, since there are V surfaces in a distance a centimetres, for the potential of CD is V, of AB it is zero, the distance between consecutive surfaces is a/V. In the second case it is a'/V'.

But if the two fields of force be the same the resultant force at each point of AB is the same in both cases and the charge on AB is the same; hence the attraction on AB will be the same. The condition for this will be that the distance between consecutive surfaces should be the same in the two cases, and for this we must have

$$\frac{a}{V} = \frac{a'}{V'},$$

or

$$\frac{V}{V'} = \frac{a}{a'}.$$

Thus we can compare the potentials of two bodies by comparing the distance between the two discs required to bring AB into the sighted position when first the one body and then the second is connected to CD.

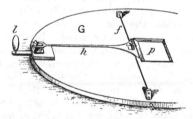

Fig. 53.

In practice it would be inconvenient to suspend the disc from a balance, and the instrument usually takes one of the two forms shewn in Figs. 53 and 54. In Fig. 53

the disc p is carried at one end of a long lever. This
is attached to a horizontal wire tightly stretched as at
f and balanced by a counterpoise. The torsion of the wire

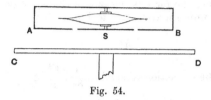

Fig. 54.

is so adjusted that in the normal position the disc p is
slightly above the guard-ring G. The arm supporting the
disc carries a light pointer which moves in front of a vertical
scale which is viewed by a lens l. On the scale are two marks
which are so adjusted that when the pointer bisects the
distance between them the disc is in its sighted position; this
is easily determined by the aid of the lens. This arrangement
is adopted in Lord Kelvin's portable electrometer.

In the form shewn in Fig. 54 the disc S is supported by
three fine springs, two are shewn in the figure, which keep it
normally slightly above the guard-ring AB; the experiment
consists as before in bringing the lower disc up until the disc
reaches the sighted position, in which as illustrated it is in
the same plane as AB. This is determined by the aid of a
lens. If it be wished to find the force exerted by the springs
this can be done by carefully loading the disc and observing
the weights required to bring it to the sighted position. This
is the arrangement in Lord Kelvin's absolute electrometer.

61. Electrostatic Measurement of Potential.
We can determine the relation between the potential of CD
and the force of attraction in the absolute electrometer thus.
If R is the resultant force near a charged conductor, and σ the
surface density, then it follows from the mathematical theory
that the pull on the surface per unit of area of the surface is
$\frac{1}{2}R\sigma$[1]. Now in the case in point the pull is in the same
direction at each point. Thus the attraction on a surface of
area S is $\frac{1}{2}R . \sigma . S$.

[1] *Elements of Electricity and Magnetism*, J. J. Thomson, § 37.

But by Coulomb's law $R = 4\pi\sigma$. Thus,

$$\text{the attraction} = \frac{1}{8\pi} R^2 S = 2\pi\sigma^2 S.$$

Moreover since the force is the rate of change of potential and since this is uniform and equal to V/a we have

$$R^2 = \frac{V^2}{a^2}.$$

Hence the attraction is equal to $\frac{1}{8\pi} \cdot \frac{V^2}{a^2} \cdot S$.

Now let M be the mass added in the experiment with the balance or the mass required to bring the disc to the sighted position in an experiment with the springs, then it is clear that the attraction is equal to the weight of the mass M.

Hence the attraction is Mg.

Thus
$$Mg = \frac{1}{8\pi} \frac{V^2}{a^2} \cdot S,$$

or
$$V = a \sqrt{\left(\frac{8\pi Mg}{S}\right)}.$$

In this expression a is measured in centimetres, S in square centimetres, M in grammes, and g in centimetres per second per second, its value being very nearly 981.

62. The Quadrant Electrometer. In this instrument, shewn diagrammatically in Fig. 55, a light disc of aluminium known as the needle is supported in a horizontal position by a wire, any force tending to displace it from its position of equilibrium is opposed by the torsion of the wire. Four metal quadrants A, B, C, D each carried by an insulating stem are placed horizontally below the needle, which lies symmetrically with regard to them. The quadrants A and D are connected together, as likewise are B and C. Suppose now that the needle is electrified. If the four quadrants be at the same potential its equilibrium is not affected; if, however, there be a difference of potential between A and B, each end of the needle, if its own potential be positive, is repelled from the quadrants at high potential towards those at low, the needle therefore is twisted until this repulsion is balanced by the torsion of the wire, and the angle through which it is twisted can be shewn to be proportional to the difference of potential between the quadrants.

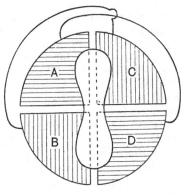

Fig. 55.

In practice the quadrants are doubled, a set being placed over as well as below the needle, the corresponding quadrants being connected together, so that the needle is practically entirely surrounded and hangs in a kind of hollow box.

To measure the deflexion of the needle a mirror is attached to it, and reflects a spot of light on to a scale. In order to use the instrument the four quadrants are connected together, and adjusted until the needle hangs symmetrically, the connexion is then removed, and the position of the spot is observed; one pair of quadrants is then usually put to earth, the other is connected to the body whose potential is required, and the deflexion of the spot is measured. This is approximately proportional to the potential.

If instead of connecting one pair of quadrants to the earth we connect it to a body at potential V' and if V be the potential of the other pair, V_0 that of the needle,—in practice

V_0 is large compared to V or V',—then the deflexion is approximately proportional[1] to $V_0 (V - V')$.

Thus it is necessary that the potential of the needle should remain the same throughout the observations; now the capacity of the needle is small, and therefore a small leakage may make a considerable change in its potential. To overcome this a piece of platinum wire hangs from the needle into a glass vessel of strong sulphuric acid placed beneath the quadrants. The vessel is coated outside with tinfoil, which is earthed by contact with the case of the instrument, and thus forms a Leyden jar of large capacity with the sulphuric acid for the inner coating. By this means the rate of fall of potential of the needle is checked.

The instrument is covered with a glass case, and the sulphuric acid serves in addition to keep the air within the case free from moisture.

Fig. 56 shews a simple form of quadrant electrometer.

EXPERIMENT 15. *To shew that the potential difference between the quadrants of a quadrant electrometer is proportional to the deflexion.*

Take three or four condensers of equal capacity,—some Leyden jars made of the same kind of glass, of the same size and thickness will be convenient,—insulate all but the last and connect them in "cascade[2]" as shewn in Fig. 57, where we suppose there are four condensers.

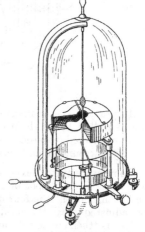

Fig. 56.

[1] The more complete formula is that the deflexion
$$= k (V - V') \{V_0 - \tfrac{1}{2} (V + V')\}$$
where k is a constant depending on the instrument. From this the result in the text follows by supposing V and V' small compared with V_0.

[2] See Section 45. Instead of the Leyden jars a number of brass plates of equal area placed in a pile at equal distances apart and insulated by small pieces of ebonite may be used.

Let A_1, A_2, A_3, A_4 be the successive inner coatings, B_1, B_2, B_3, B_4 the outer coatings; A_2 is connected to B_1, A_3 to B_2, and so on. Charge A_1 to a potential V; the potential of B_4 is zero, for

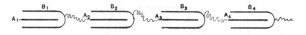

Fig. 57.

it is earth-connected, and the fall of potential from V to nothing will be equally divided among the four condensers. This can be shewn by experiment[1], for if we connect the two coatings of each condenser in turn to a quadrant electrometer we find we obtain the same deflexion.

Thus there is a difference of potential $V/4$ between the coatings of each condenser.

Thus between A_1 and B_1 the potential difference is $V/4$.

    Between A_1 and B_2 it is $2V/4$.

     ,, A_1 and B_3 it is $3V/4$.

     ,, A_1 and B_4 it is $4V/4$.

Now connect A_1 to one pair of quadrants and connect in turn to the other pair B_1, B_2, B_3, B_4 noting the deflexion in each case. They will be found to be respectively proportional to 1, 2, 3, and 4. Hence the deflexions observed are proportional to the potential differences between the quadrants. Thus we may use the quadrant electrometer to compare differences of potential by comparing the deflexions produced.

Moreover, if we can determine in any way the potential difference which produces a given deflexion we can find the constant of the instrument and so use it to measure any other difference of potential.

63. Electrostatic and Multicellular Voltmeters.
The quadrant electrometer was invented by Lord Kelvin, who has given it various forms. In the electrostatic voltmeter[2]

[1] This follows at once from the theory given in Section 45, if we suppose all the condensers to have equal capacity.

[2] A volt is the name given to the unit difference of potential, and a voltmeter is an instrument for measuring volts.

(Fig. 58) two opposite pairs of quadrants are removed, the
axis round which the needle turns is horizontal; the lower
end of the needle carries an adjustable weight, and the needle

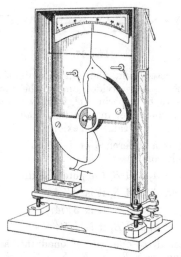

Fig. 58.

is connected to the quadrants. When the instrument is
electrified the needle and quadrants are raised to the same
potential and the needle is repelled from the quadrants. The
motion of the needle is indicated by a pointer which moves
over a graduated scale. In this case the force on the needle
is proportional to the square of the potential; this follows at
once if we put $V_0 = V$, $V' = 0$ in the formula in the note to
Section 62, which then gives us the deflexion as equal to $\frac{1}{2} k V^2$.

We can also use the electrostatic voltmeter to measure the
difference of potential between two bodies by connecting one
body to the needle, the other to the quadrants.

In the multicellular voltmeter the sensitiveness of the
instrument is increased by attaching to the same vertical
wire a number of needles each of which hangs in a horizontal
position. These are attracted by a series of quadrant-shaped

discs placed between the needles as shewn in Fig. 59.   In
this instrument also, which is used for measuring high potential
differences, it is usual to connect one of the two points between
which the potential is required to the needle, while the other
is joined to the quadrants.

The deflexion is then proportional to the square of the
difference of potential.

**64. Potential at a point
in the air.**   It is clear that an
insulated conductor need not be
at the same potential as the air
round it ; if however there be a
difference of potential between a
conductor and the air, and we
can arrange that a stream of
small conducting particles shall
leave the conductor, since positive electricity tends to pass from
places of high to places of low
potential, the conductor, if at a
higher potential than the air, will
discharge positive electricity with
the stream of particles, and conversely if it is at a lower potential
it will discharge negative electricity ; in either case the effect

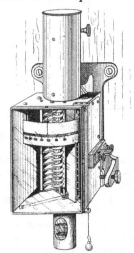

Fig. 59.

will be to bring the conductor to the potential of the air
in its neighbourhood.   If then the conductor be connected to
an electrometer the readings of the instrument will measure
the potential of the air at the point where the stream of
particles breaks up.

In some cases the stream of particles is a fine jet of water
issuing from a long nozzle attached to an insulated vessel, the
vessel is connected with an electrometer, which thus measures
the potential of the air ; if it is wished to obtain a continuous
record of the changes, the quadrant electrometer is employed,
and the spot of light from the mirror is focussed on to a sheet
of sensitive paper wound on a drum, which is made to move
by clockwork.

The motion of the spot is parallel to the axis of the drum, at right angles therefore to the motion of the paper.

The trace of the spot when developed forms a continuous curve, which gives the variations of the potential. Such a curve obtained at the National Physical Laboratory is shewn in Fig. 60.

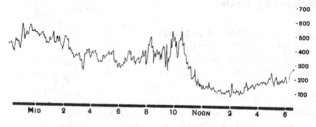

Fig. 60.

In another arrangement due also to Lord Kelvin the necessary stream of particles is afforded by the smoke from a slow-burning match; the match is held in an insulating handle and connected to the electrometer, and the potential measured by reading the instrument. The portable electrometer (Fig. 61) is usually employed for this purpose.

Very large variations of potential are observed, during a day the changes at a given point may amount to 500 or even 1000 volts.

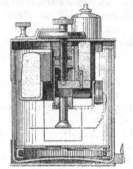

Fig. 61.

EXAMPLES ON ELECTROSTATICS.

1. A rod is brought near to a magnetic needle, and the latter is thereby deflected. By what further experiment would you determine whether the observed action is magnetic or electrical?

2. A small positively charged sphere is placed near the end of a small insulated uncharged cylinder, and both are placed near the centre of a large spherical metallic shell connected with the earth. Describe in general terms the distribution of the charges on the bodies, and sketch the lines of force of the system.

3. The end of a wire connected to a gold-leaf electroscope is put through a hole in the side of a hollow charged conductor. Describe and explain what happens when (1) it is held there without being allowed to touch the conductor, (2) it is withdrawn, (3) it is again inserted and allowed to touch the inside of the conductor, (4) it is withdrawn, (5) it is made to touch the outside of the conductor.

4. A gold-leaf electroscope is put inside a tin can, which is hung up by silk cords so as to be insulated. On holding a strongly electrified glass rod below the can no divergence of the leaves takes place; but on touching the cap of the electroscope with the fingers (without touching the can) the leaves diverge. Explain these results.

5. Two exactly similar gold-leaf electroscopes have their caps connected by a wire and a positively charged body is brought near the cap of one of them. Compare and explain their indications.

If the wire be now removed by means of an insulating handle and then the charged body also removed, what effects will be observed in the electroscopes?

6. A hollow cylinder is charged positively and insulated. An insulated metal funnel is placed with its nozzle well within the cylinder and water is allowed to drop through the funnel into an insulated vessel placed below the cylinder. Shew that the vessel will become negatively electrified.

7. A large box is coated with tinfoil and insulated. A wire insulated from the walls of the box and connected to the earth passes to the inside. How can a man inside the box, if he has the necessary apparatus, determine whether the box be charged or not?

8. Would an electrophorus work better when the cake of pitch or resin is placed on a conducting plate or when it is in contact with non-conductors only?

9. How could you give to a hollow vessel B double the charge of a small conductor A without discharging A?

10. If a charged ebonite rod be placed in contact with the knob of an electroscope the leaves diverge and on its removal they partially collapse. Why is this?

11. If an electroscope be charged and a body with a big charge be brought near it, state and explain what will be the indications of the electroscope according as the charges are of the same or of opposite sign?

12. If a small spherical conductor with a strong positive charge is gradually brought near to a large spherical conductor with a weak positive charge repulsion followed by attraction is experienced. Explain this.

13. A hollow insulated conductor is electrified positively. A small insulated conducting sphere, connected by a fine wire with a gold-leaf electroscope which is negatively electrified, is placed (a) near the outside; (b) touching the outside; (c) inside; (d) touching the inside of the hollow conductor. Describe and explain in each case the result, on the electroscope, of the experiment.

14. A ball held by a damp silk thread is introduced inside a charged hollow conductor, and made to touch the side. It is brought out and presented to an electroscope which indicates a charge. Explain this fact.

15. Two equal small spheres are charged when in contact and then placed at a distance of 1 metre apart. They are found to repel each other with a force of 900 dynes. Find the charge on either.

16. What charge shared between two equal spheres 5 cm. apart will cause them to repel each other with a force of 81 dynes?

17. Two small equal metal spheres are placed 5 cm. apart. What is the force between them if one has a charge of $+5$ units and the other -10 units? What does the force become if they are connected for a moment by a wire? Would there be any force if one of them were connected with the earth and the other charged?

18. Two small spheres a and b charged with the same quantity of positive electricity are placed at a distance of 1 metre. Where should a sphere c holding twice the amount of electricity be placed so that the electrical forces on b may be in equilibrium (1) when c is charged positively, (2) when c is charged negatively?

19. Sketch the lines of force and the equipotential surfaces due to two small spheres placed 1 metre apart (a) when charged with quantities $+400$ and $+100$ units respectively, (b) when charged with quantities $+400$ and -100 respectively; and find in each case the position of the point at which the force is zero.

20. Two small pith balls, each weighing one gram, are suspended from the same point by silk fibres each 12 cm. long, and of negligible mass. They are then equally electrified so that each string makes an angle of 45° with the vertical. What charge must each ball possess?

21. It is required to determine which of two electrical conductors has the larger capacity. Explain how you would decide the question if you were provided with a sensitive gold-leaf electroscope and an electrophorus.

22. The knob of an electroscope is connected with an insulated metal plate : this is then charged with electricity: a metal plate, held in the hand, is brought up opposite the insulated plate; state and explain what will be the indications of the electroscope.

23. Three Leyden jars whose capacities are C_1, C_2, C_3 are insulated, except that the outer coating of the third jar is earthed; the outer coating of the first is connected to the inner coating of the second, and the outer coating of the second to the inner coating of the third. A charge Q is given to the inner coating of the first. What are the charges on the inner coatings of the other two, and what is the difference of potential between the inner coating of the first and the outer coating of the third?

24. One plate of a plate condenser is connected to a gold-leaf electroscope and the following operations are performed: (1) the other plate is charged positively; (2) the electroscope is momentarily put to earth and then again insulated; (3) the distance between the plates is increased; (4) a plate of glass is inserted between the plates. Describe and explain the effect on the electroscope in each case.

25. Two metal spheres, one six times the diameter of the other, are connected by a long thin wire and electrified. Compare their electric charges, potentials, densities and energies.

26. Find the work spent in giving a sphere of 10 cm. radius a charge of 50 units.

27. The capacity of a conductor is 20 c. g. s. units. What must be its charge in order that its energy may be 1000 ergs?

28. How is the electrical energy of a charged air condenser affected if a cake of sulphur is introduced between the plates?

29. A body of capacity C is charged to a potential V. What change in energy occurs when its charge is shared with another body of capacity C'?

30. An insulated charged sphere, 3 cm. in radius, shares its charge with another, 6 cm. in radius; what is the relation between the energy of the charged body before and after it has shared its charge?

31. The opposite plates of an air-condenser are connected with a quadrant electrometer, which shews a small deflexion θ when the condenser is charged by means of a Daniell's cell. What will be the deflexion if the air be replaced by a slab of paraffin (specific inductive capacity $= 2$), (i) before, and (ii) after the cell is disconnected from the condenser?

32. Three insulated conducting spheres, whose radii are respectively 2, 3 and 4 cm., are so charged that their respective potentials are 6, 4 and 3 units of potential. They are then connected by wires of negligible capacity. Find the total quantity of charge, the total capacity, and the final common potential of the system.

33. An insulated cylindrical sheet of tinfoil is electrified so that two pith-balls, suspended from the foil by cotton threads, diverge widely. The foil is then unrolled (still remaining insulated) and it is found that the angle of divergence is diminished. Explain this result.

34. An air-condenser is charged and connected to an electrometer; the distance between the plates of a condenser is measured; a plate of ebonite of known thickness is inserted between the plates and their distance is adjusted so that the electrometer deflexion does not change. Shew how to determine hence the specific inductive capacity of the ebonite.

35. Find the capacity of a condenser composed of a pair of circular plates 18 cm. diameter and 1 mm. apart, with air as the dielectric.

36. Find the capacity of a spherical condenser the radii of whose surfaces are 15·9 and 16·1 cm. respectively. What would be the effect of replacing the air between the spheres with turpentine?

37. A condenser is composed of two square plates each 10 cm. in side; the plates are 1 mm. apart. It is found that 500 ergs are needed to charge it to a constant difference of potential. Find the difference of potential between the plates and the charge on either plate.

38. A Leyden jar is charged to potential 1200. It is then connected to another jar with twice the area of coating but of the same thickness and previously uncharged. What is the common potential? If the second jar had been previously charged to potential 600 what would be the common potential?

39. The diameter of a Leyden jar is 12 cm., the height of the metal coating 15 cm., the thickness of the glass 2 mm., and its specific inductive capacity 3·1. Calculate the electrostatic capacity, and find the work done in charging it to a potential V.

40. Calculate the total heat developed by the discharge of a condenser consisting of two concentric spheres separated by paraffin when the charge on one sphere is 100 electrostatic units. The radii of the spheres are 10 and 12 cm. and the dielectric constant of paraffin is 2.

41. An attracted disc electrometer is immersed in an insulating oil with specific inductive capacity 2. The area of the attracted disc is 50 sq. cm., and its distance from the fixed plate is 0·5 cm. If the electric pull on the disc is 500 dynes, find its difference of potential from that of the fixed plate.

CHAPTER VII.

MAGNETIC ATTRACTION AND REPULSION.

65. Natural Magnets. It was known to the ancients that certain black stones—iron ores, which were found commonly at Magnesia in Asia Minor—possessed the power of attracting to themselves small pieces of iron. Such stones were called magnets. Many centuries later it was found that a magnet when suspended freely by a thread tended always to set itself in one definite position[1]. One end was observed always to point north, the other south. Hence the magnet acquired the name of lodestone or leading stone. The only force we know of acting between the earth and most bodies is that of gravitation; in the case of a magnet an additional magnetic force comes into play and the position the magnet assumes depends in part on this force.

These natural magnets are composed of an oxide of iron, having the chemical composition Fe_3O_4, and are found in many parts of the world besides Magnesia.

After a time it was discovered that these magnetic properties could be communicated to a piece of iron which was stroked by a natural magnet, and still later it was shewn that hardened steel when stroked retained its magnetic properties more permanently than soft iron.

Dr Gilbert, who in 1600 published a book, *de Magnête*, about magnetism, added much to our knowledge by his discoveries. He shewed for example that if a bar of steel

[1] It is said that the Chinese were acquainted with this property at a very early date.

whose length is considerable in comparison with its width be magnetised, the magnetic attraction is exerted most markedly by the ends of the bar, and hence he introduced the idea of the magnetic poles of a magnet, two points towards which he supposed the attractive forces to act. The line joining these two poles he called the axis of the magnet.

Moreover he proved that this magnetic attraction was exerted across almost all kinds of bodies practically unchanged; a sheet of wood or brass or lead may be inserted between the magnet and the soft iron without modifying the effect. The magnet may be enclosed in a vessel or sealed up in a glass tube containing air or any gas or exhausted, as the case may be, the force remains unaltered. A screen of iron however will affect the action of the force. A magnet inside a very thick hollow shell of soft iron exerts no sensible force outside the shell and is unaffected by external magnets.

It is known now that an actual magnet does not possess "poles," two centres, that is, of attraction; the axis of a magnet however is a term which we shall use continually, and of which we can give a definition quite independent of the idea of poles; at the same time in dealing with long, thin magnets it is convenient to use the term pole as indicating the end of the magnet, and as we shall see the effects of the magnet can be calculated approximately as though the poles were real centres of attraction.

66. Artificial Magnets. A steel magnet may take many forms; a long rectangular bar, such as is shewn in Fig. 62, is

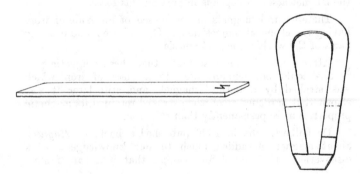

Fig. 62.   Fig. 63.

known as a bar magnet; sometimes the bar is bent into the form of a horse-shoe (Fig. 63). In a compass needle (Fig. 64) a thin strip of steel in the form of a very elongated lozenge is

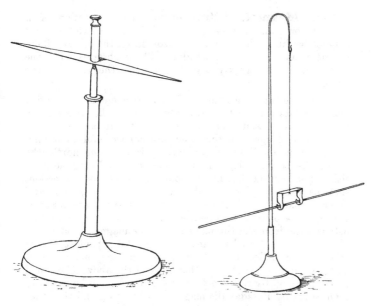

Fig. 64.   Fig. 65.

employed. This carries at its centre a small cup of glass or agate by which it can be supported on a sharp point. For many of our experiments a thin knitting needle will be found useful; this can if necessary be suspended in a small stirrup of brass or copper wire by means of a silk fibre (Fig. 65).

Fig. 65 a.

A spherical ended magnet (Fig. 65 a) as constructed by Robeson may conveniently be made out of two ½ inch steel bicycle balls connected by a piece of knitting needle six inches in length; in such a magnet the poles coincide very nearly with the centres of the balls.

67. Magnetic Attractions. The attraction of a magnet for iron may be illustrated in many ways. If a magnet be dipped into a vessel of iron filings the filings adhere to the magnet, and in general it will be observed that the filings are most thickly distributed near the ends of the magnet.

Take a compass needle or, if more convenient, suspend a magnet in its stirrup by a silk fibre. Note the direction in which it places itself, and observe that after it is disturbed it comes to rest again in the same position as before, one end, which we will call the north pole, points to the north, the other, the south pole, points to the south. Bring near to either pole of the magnet various substances in succession, such as rods of glass, wood, copper, brass, lead, etc., and observe that none of these appreciably disturb the magnet. Now bring a piece of iron near either pole; there is attraction between the iron and the magnet, and the magnet is deflected. The iron is said to be a magnetic substance; most of the others were called by Gilbert non-magnetic; more careful observation however shewed that some of the above-mentioned bodies are slightly repelled by the magnet. Such bodies have been called by Faraday dia-magnetic, while iron and the other magnetic bodies which are attracted were named para-magnetic. Other substances which are attracted by the magnet, and are therefore para-magnetic, are nickel and cobalt, but the attraction exerted on these bodies is very much less than in the case of iron. In fact for our purposes we may look upon iron—including steel—as magnetic, and other substances as non-magnetic. We must carefully distinguish between a magnet which sets in a definite position and attracts iron filings and a magnetic substance, such as iron, which is attracted by a magnet, but which in its normal condition does not attract other magnetic material. A magnetic substance can be magnetised; a magnet is a piece of such a substance which has been magnetised. We have tacitly assumed above that the iron rod is not itself a magnet; it remains to describe what happens when we bring a second magnet near the first.

EXPERIMENT 16. *To examine the forces between two magnetic poles.*

Take two compass needles and support them at some distance apart.

Notice that the needles point in the same direction, approximately north and south, and mark the north or north-pointing end of each.

Dismount one needle and bring each end in turn to the two ends of the second needle, observing the results. It will be found that when the two like ends are close together there is repulsion, when the two unlike ends are brought near there is attraction. The results may be summed up thus:

| | Effect on Compass Needle | |
	North Pole	South Pole
North Pole of dismounted needle	Repulsive	Attractive
South Pole of dismounted needle	Attractive	Repulsive

Hence we arrive at the result that there is repulsion between two like, attraction between two unlike magnetic poles.

We can vary these experiments by putting plates of various materials between the magnetic poles. We shall find that unless the material interposed contains iron[1] the forces between the poles are the same as previously.

Thus we see that on a given magnetic pole—the end of one magnet—opposite effects are produced by the two ends of a second magnet. We may compare this with the opposite

---

[1] There are some few other substances besides iron, e.g. nickel and cobalt, which will modify the force very slightly, but so slightly that much more delicate experiments than those described are required for its detection.

effects produced on a positively electrified body by two conductors, one charged positively, the other negatively, and may
ascribe opposite qualities—positive and negative respectively—
to the two ends of a magnet. The north-pointing end is
called the positive end; it is described as being positively
magnetised, or as possessing a positive charge of magnetism;
it is called a positive pole; the south-pointing rod is negatively
magnetised, or possesses a negative charge; it is a negative
pole. As in the case of electricity, the choice of sign is
conventional.

When one magnet is brought near a second the resulting action is complex; in the first place the magnet does
not strictly possess poles, we cannot regard one point as
a centre of attraction, another as a centre of repulsion, and
deduce all the forces from the action of these two centres;
however with a long, thin magnet we may very approximately
make this assumption, but even in this case we have four
forces to consider.

For let NS, $N'S'$ (Fig. 66) represent the two magnets,
then there are forces of repulsion along NN' and SS' respectively, and of attraction along NS' and $N'S$. If however

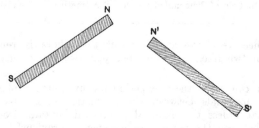

Fig. 66.

NN' is small and the other distances considerable, the resulting
action will be mainly due to the repulsion along NN'.

Again, although it is convenient to speak of a magnetic pole and of
the force exerted by such a pole we must remember that we can never
obtain a single pole. Any magnet has always two poles; if we have an
electrical conductor positively electrified at one end, negatively at the

other, and divide it into two portions we can separate the positive and the negative electrification, we cannot do this with a magnet; if we take a magnet NS and cut it in two in the middle, each of the two halves remains a magnet with a north and a south pole.

68. Magnetic Induction. A magnetic substance may be temporarily magnetised by the presence of a magnet. Take a rod of soft iron and dip it into a vessel of iron filings, on removing it few[1] if any of the filings adhere to the iron. Repeat the experiment, holding one end of a powerful bar magnet near to the upper end of the iron rod. The iron filings adhere plentifully to the rod, but many drop off when the magnet is removed. The rod has been magnetised by induction; the effect is very much increased if the magnet be allowed to touch the iron rod.

Thus a whole string of iron tacks can hang suspended, as shewn in Fig. 67, from one end of a magnet; if the uppermost tack be separated from the magnet the chain falls to pieces.

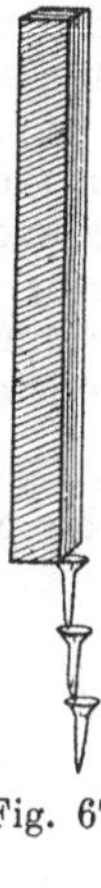

Again, when a rod is magnetised by induction the end nearest the inducing pole is of opposite sign to that pole. To shew this bring one end of a rod of soft iron near the north end of a compass needle, it is attracted to the iron, then bring the north pole of a strong magnet near the further end of the iron, the compass needle is now repelled; the end of the iron rod which originally attracted the north end of the compass has been positively magnetised by induction and now repels it. This enables us to account in some measure for the attraction between the magnet and the iron. For let NS (Fig. 68) be the magnet and AB a piece of iron, the end A being near to N and B at some distance. The rod AB is magnetised by induction, A becoming a south end, B a north. The south end A is then attracted by the north end N of the original magnet.

Fig. 67.

69. Magnetic Field. Thus a magnet exerts magnetic force on any other magnetic body in its neighbourhood; the

---

[1] If the iron happens to have been recently magnetised a few of the filings may adhere to it.

space round the magnet throughout which this force is exerted
is called its magnetic field, and at any point of the field there
exists magnetic force of a definite amount acting in a definite
direction.

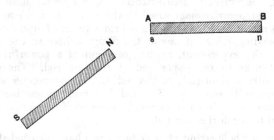

Fig. 68.

Since a magnet freely suspended at any point on the
earth's surface sets in a definite direction, its north pole
always pointing approximately to the north pole of the earth,
the magnet is in a field of force due to the earth; the earth
itself exerts magnetic force. Experiments will be described
later by which the strength and direction of this field can be
determined. For the present it is sufficient to bear in mind
its existence.

70. Magnetic Lines of Force. Suppose now that
at the point P (Fig. 69) magnetic force is acting in the

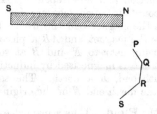

Fig. 69.

direction PQ, that is to say, a north or positive pole placed
at P is urged in the direction PQ; if we move on to Q, a

point near to P lying on the line of action of the force at P, the direction of the force will in general change; let it become QR. At R a point near Q the direction will again have altered to RS suppose. We thus have a line $PQRS...$ made up of a series of short straight pieces PQ, QR, RS, etc. which has the property that these straight pieces respectively each give the direction of the magnetic force at the respective points P, Q, R. Now if the lengths of these pieces be very much reduced we arrive finally at a continuous curve $PQRS$ (Fig. 70), and this curve has the property that its direction at each point of its length

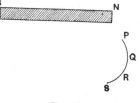

Fig. 70.

gives the direction of the resultant magnetic force at that point; it is called a line of magnetic force.

We arrive thus at the following

DEFINITION. **A line of Magnetic Force** *is a line whose direction at each point of its length gives the direction of the magnetic force at that point.*

Thus we may picture to ourselves the magnetic field as permeated by the lines of magnetic force; if in any case we can map out the lines of force we can calculate the distribution of the force in the field.

The above definition and explanation of a line of magnetic force should be compared with those given in Section 20 of lines of electrical force.

PROPOSITION 4. *A small magnet placed in a magnetic field sets itself with its axis along the line of force through its centre.*

Suppose we place a small magnet NS (Fig. 71) in the field and let it be free to move about an axis through its centre O. Let OP be the direction of the magnetic force at O, the direction that is in which a north pole at O would be urged. The north end N of the small magnet is pushed in the direction OP, the south end S pulled in the opposite direction; thus NS is turned until its axis points in the direction of the force OP; the direction then in which the axis of the small magnet rests is that of the line of force through its centre.

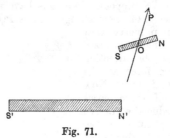

Fig. 71.

It is not necessary for the success of this method of determining the direction of a line of force that NS should be a magnet, if it be an elongated piece of soft iron placed as in Fig. 71, then N becomes by induction a north pole, S a south pole under the action of the force, and the results are the same, the direction in which the soft iron sets is that of the line of force. There is this difference in the two cases; with the magnet we can tell the direction of the line of force, it will be from the south pole S to the north pole N; with the piece of soft iron either end may have become a north pole by induction, we have nothing to tell us which it is.

71. Tracing Lines of Force. We may employ either of these methods to trace the lines of force in a magnetic field.

EXPERIMENT 17. *To trace the Lines of Force in a magnetic field by the use of a small compass-needle.*

Fasten a sheet of drawing paper on a horizontal drawing board, and place a small compass on the board, removing all other magnets from the neighbourhood. Make a dot with a pencil opposite each end of the compass-needle. Move the compass so that its south pole is over the dot which was opposite its north pole, and make a dot opposite its north pole.

A small compass-needle pivoted in a brass cell between two discs of glass is convenient for this experiment; the dots can be seen through the glass.   Continue this process and thus find a series of positions indicating the direction of the resultant force at the successive points.   Draw a line through all the dots; this will be a line of force.   In the same way draw a second line of force on the paper.   It will be found that the two lines are straight and parallel, they indicate the direction in which the magnetic force due to the earth acts on the magnet, and run north and south.   Now take a bar magnet and place it on the paper.   Draw a line round the

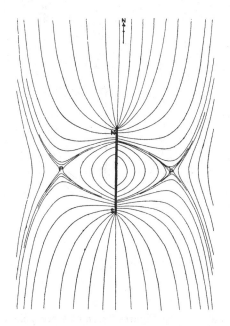

Fig. 72.

magnet, and mark off a number of points on this line; from these points draw a series of lines of force, using the compass as before.

Repeat this for different positions of the magnet. The form of the lines obtained which are due to the combination of the force due to the magnet and that due to the earth will depend on the position of the magnet, and on its strength. If the axis lies north and south, with the north pole to the north, they will be as in Fig. 72, if the north pole points to the south the lines of force will be as in Fig. 73. In Figure 74 the axis of the magnet is inclined to the north and south line.

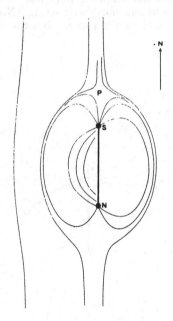

Fig. 73.

The points P, P in Figures 72 and 74 are points at which the force due to the magnet exactly balances that due to the earth. They are thus points of zero force.

The lines of force due to a combination of magnets can be traced in a similar manner.

The general forms of the lines of force can be more easily observed by the use of iron filings. The magnet or magnets whose field is to be explored are placed on the table. Over

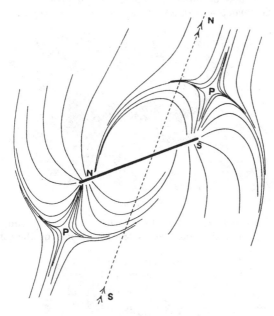

Fig. 74.

them is placed a thin plate of glass, and on the glass a sheet of paper. Iron filings are dusted on to the paper through a sieve, and the glass plate is tapped so as to aid the filings to take up their positions under the magnetic forces.

Figures[1] 75 a to e shew the forms of the lines of force due to a single magnet, and also to two magnets placed in the positions indicated.

[1] These figures are taken by the author's kind permission from Watson's *Text-book of Physics*. Longmans.

72. Tensions and Pressures in a Magnetic Field. In dealing with electrostatic action we have seen we can explain the phenomena by supposing a tension to exist along the lines of force, combined with a pressure at right angles to them; we can apply the same idea to magnetism.

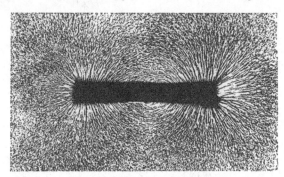

Fig. 75 a.

In Figures 75 b, c, and e lines of force are seen joining the north pole of either magnet to the south pole of the other;

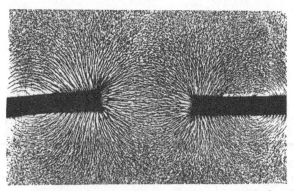

Fig. 75 b.

along these lines there is attraction; if we follow the lines from the north pole of one magnet which start in the direction of the north pole of the other, they are apparently deflected from their course by the lines which issue from the second north pole, and the two series of lines run off together in

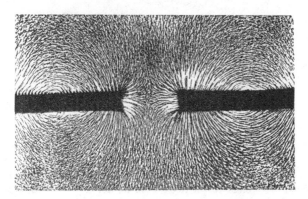

Fig. 75 c.

more or less parallel directions : between these two lines there is repulsion ; the two like poles repel.

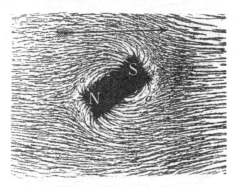

Fig. 75 d.

We may also examine by this method the consequences of placing a magnet in a uniform field. This is shewn in

Fig. 72, in which near the boundaries of the figure the magnetic lines run uniformly from left to right; the notion

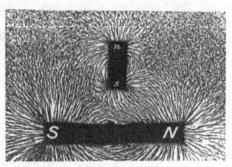

Fig. 75 e.

of tension along the lines will explain how it is that the magnet *NS* tends to set itself with its north pole to the right of the figure.

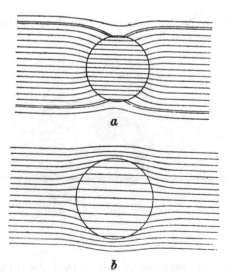

a

b

Figs. 76 a, 76 b.

If again a sphere of soft iron be placed in the field, the lines of force in its neighbourhood are no longer uniformly distributed. They crowd into the iron at one side and leave it again at the other as shewn in Fig. 76 *a*; the iron is said to have a greater permeability to the magnetic induction than the air which it displaces.

If the sphere be diamagnetic the distribution of the lines is as in Fig. 76 *b*.

If a ring of soft iron be placed in the field of a bar magnet the effect is as shewn in Fig. 77. There are practically no lines of force within the ring. This explains how it is that a

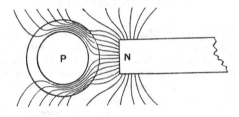

Fig. 77.

thick shell of iron acts as a magnetic shield, the lines of force from outside do not enter the air space in the interior of the shell; the magnetic force within is much less than it would be if the ring were removed.

CHAPTER VIII.

LAWS OF MAGNETIC FORCE.

73. Charges of Magnetism. We shall find it convenient to speak of a magnet as being charged with magnetism. In the case of a long thin magnet the charges—the one positive, the other negative—reside near the ends of the magnet, and may be considered as concentrated at two points, the north and south poles of the magnet respectively; the forces of attraction or repulsion observed between two magnets arise on this view from the repulsions or attractions between the charges of magnetism in the magnets. A magnetic pole is a point at which a charge of magnetism is concentrated, and the force between two poles is the force between the two charges concentrated at these poles. This charge is known as the strength of the pole.

We may also view a magnetic pole as a centre from which lines of force diverge or to which they converge, and picture to ourselves the action between the poles as due to the tensions and pressures set up in their neighbourhood by the presence of these lines of force. According to this view the strength of a pole will be proportional to the number of lines of force which diverge from it. This latter method will probably lead us further in any attempt to explain the cause of magnetic force, the former lends itself more readily to calculation.

74. Law of Magnetic Force. Coulomb was the first to determine the law of force between two magnetic poles. He shewed by means of the torsion balance that the force between two poles of strengths m, m' placed at a distance r apart is proportional to mm'/r^2.

When the torsion balance is used for magnetic experiments the light shellac needle which carries the pith balls (shewn in Fig. 51) is replaced by a long thin magnet, and the charged conductor by a second such magnet. The instrument is then used in the same way as in the electrical investigation except that a correction is required for the effect due to the earth's field. The law of force however is more accurately verified by an experiment due to Gauss described in Section 94; the interest of the torsion balance is chiefly historical.

Thus it follows from these experiments that the force between two magnetic poles m, m' follows the same law as that between two electrical charges e, e'. If we measure our various quantities in proper units we may write

$$F = \frac{mm'}{r^2} \text{ dynes.}$$

75. Unit Magnetic Pole. Suppose now we have two exactly equal poles, so that m is equal to m', and that they are placed at a distance of 1 cm. apart, so that r is equal to 1; suppose further it is found that the force of repulsion is 1 dyne, so that F is 1. Then substituting in the above equation we have

$$1 = \frac{m^2}{1^2},$$

or $$m = \pm 1.$$

Hence each pole must be of unit strength and we obtain the following

DEFINITION OF UNIT POLE. **A Magnetic Pole** *has* **Unit Strength** *when it repels an equal pole placed at a distance of 1 centimetre with a force of 1 dyne.*

Thus with this definition of unit strength we may say that there is a force of repulsion between two poles of strengths m, m' placed at a distance of r centimetres apart of mm'/r^2 dynes.

If one of the two m, m' is negative, this force of repulsion is negative, that is, the force is one of attraction.

In strictness just as in the corresponding equation in electricity a quantity K is introduced which represents the action of the dielectric medium, so here we ought to introduce a quantity—μ—to represent the magnetic action of the space between the two poles ; since however the effects of all media except iron and to a less degree nickel and cobalt are very nearly indeed the same as that of air, μ, the permeability of the medium, as it is called, is assumed to be unity and not explicitly introduced into the equation which ought more accurately to be written

$$F = \frac{1}{\mu} \frac{mm'}{r^2}.$$

The fact that μ is unity for most materials is proved by the observation that the force between two magnetic poles is not altered by placing various materials between the two. Compare Section 27.

76. Total Magnetic Charge of a Magnet. We have said that the magnetic action of a magnet may in many cases be represented as due to two opposite poles, the one placed near its north-pointing end, the other near its south-pointing end; such a magnet is called a solenoidal magnet.

It follows accurately as the result of experiments described in Section 90, that the strengths of these two poles are equal; if the north pole contains m units of magnetism the south pole contains $-m$. Thus the total quantity of magnetism in the magnet is zero.

77. Solenoidal Magnet. In dealing with a solenoidal magnet the following definitions of the terms magnetic axis and magnetic moment will be useful.

DEFINITION. *The line joining the poles of a magnet is called its* **Magnetic Axis.**

DEFINITION. *The product of the strength of either pole of a magnet into the distance between the poles is called the* **Magnetic Moment of the Magnet.**

DEFINITION. *The ratio of the magnetic moment of a magnet to its volume is known as the* **Intensity of Magnetisation** *of the magnet.*

More general definitions of the first two of these quantities applicable to all magnets will be found in Section 90.

78. Resultant Magnetic Force. Intensity of the Field. If a unit positive pole be placed at any point of a magnetic field it will experience a certain force. This force is known as the magnetic intensity of the field or the resultant magnetic force at that point of the field. Let it be R dynes. Then a pole of strength m' placed at that point will be acted on by a force of Rm' dynes.

If for example the field is due to a single pole of strength m at a distance of r centimetres, since the force between the two poles m, m' is mm'/r^2, it is clear that R is equal to m/r^2, or the intensity of the field at a distance r from a pole m is m/r^2.

We may compare these results with those at which we have already arrived in the case of a charge of electricity in Section 29.

79. Magnetic Potential. In dealing with electrostatics we have explained what is meant by the electrical potential and how it is measured; similarly in magnetism we have to consider the magnetic potential or magnetic pressure at a point, this is measured by the work done in bringing a unit magnetic pole up to the point.

We thus have the following definitions of Resultant Magnetic Force and of Magnetic Potential.

DEFINITION. *The* Resultant Magnetic Force *at a point is the force on a unit magnetic pole placed at the point.*

DEFINITION. *The* Magnetic Potential *at a point is the work done in bringing a unit magnetic pole from beyond the boundaries of the field up to the point.*

These definitions should be compared with the corresponding ones in Electrostatics (Section 31).

CHAPTER IX.

EXPERIMENTS WITH MAGNETS.

80. Methods of Magnetisation. We have seen that a piece of iron or steel can be magnetised by contact with another magnet, and that in the case of a steel bar the magnetism so produced is in great measure permanent. Powerful magnets are now generally produced by the action of an electric current in a manner which will be described later (Electromagnetism). The following experiments illustrate some of the older methods of making a magnet from a piece of steel.

EXPERIMENT 18. *To magnetise a piece of steel.*

(i) *By single touch.* Verify that the piece of steel which may conveniently be a piece of a knitting-needle is unmagnetised. Take a bar magnet and determine by the aid of a compass-needle as in Experiment 17, which is its north end.

Draw the north pole of the bar magnet from one end A to the other end B of the piece of steel, Figure 78. Repeat this several times, beginning each time at the same end. Present one end of the piece of steel to

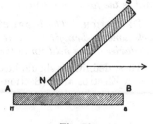

Fig. 78.

the compass-needle and observe that it is magnetised, and that the end A from which the north pole was drawn is its north pole, while B is its south pole.

(ii) *By divided touch.* For this two bar magnets of equal strength are required. Put the piece of steel on the table. Place the north pole of one bar magnet in contact with the south pole of the other, on the centre of the piece of steel. Draw the two magnets apart to the ends A, B of the (Fig. 79)

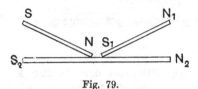

Fig. 79.

piece of steel. Repeat this several times. The piece of steel is thus magnetised; on presenting it to the compass-needle it will be found that the end A to which the north pole of one of the magnets was drawn has become a south pole, while the end B to which the south pole was drawn has become a north pole.

The effect in this case is increased by allowing the ends A, B of the bar which is to be magnetised, to rest on the north and south poles respectively of two other equally strong magnets as shewn in Figure 80.

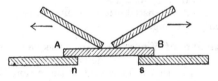

Fig. 80.

81. Magnetic Batteries. In the methods of magnetisation just described it is chiefly the outer layers of the steel that are affected. More powerful magnets are sometimes constructed by taking a number of plates of thin steel, magnetising each separately, and then building them up into a permanent magnet. In some cases two pieces of soft iron are placed as shewn in Fig. 81, over each end of the bar; these become magnetised by induction and serve to keep the ends of the steel plates together.

Another form of magnet is that used in the Kew pattern instruments for measuring the strength of the magnetic force due to the earth; in these instruments the magnets are hollow steel cylinders.

Fig. 81.

82. Demagnetisation due to the Ends. In the case of a compound magnet such as that just described, the magnetic moment of the whole is far from being the sum of the moments of its parts; for each bar acts inductively on its neighbours, and developes in them a magnetisation opposite to its own; by this action the resultant magnetism is reduced; a similar action occurs in any magnet, for let NS, Fig. 82, be a magnet and P any point in it, then at P there will be a magnetic force acting from N towards S tending to produce by induction a magnet with its north pole towards S, and its

Fig. 82.

south towards N, a distribution of magnetism opposite to that originally existing.

83. Action of Keepers. In consequence of this demagnetising action there is a tendency for any magnet to lose its magnetism, and this is increased if the magnet be subject to a jar or shock of any kind.

To reduce the action, permanent magnets are often fitted with a keeper, a bar of soft iron placed so as to connect the poles. Thus in the case of a bar magnet, two magnets are kept together with their poles arranged as in Fig. 83; the two poles N, S' are connected by one piece of soft iron AB and the two N', S by a second piece CD. By this means the demagnetising force in the magnets is reduced, for the north pole N induces a south pole at A, and the effect of this at any point P of the magnet is opposite and nearly equal to that of the original north pole.

In the case of a horse-shoe magnet a single piece of soft iron is placed across the poles, and the same effect is produced.

We have already seen that when a piece of soft iron is placed in a magnetic field the lines of force are concentrated through the iron. When a single bar magnet is alone lines of force diverge from its north pole, and pass round through the air to its south pole, many however return through the magnet itself; the demagnetising action is due to these. When a second bar magnet is placed near as in Fig. 83, the poles being

Fig. 83.

reversed in the two, even before the keepers are put on many lines of force pass from N to S' across the air gap and the number through the magnet from N to S is reduced; by placing the keepers AB and CD in position practically all the lines are made to flow through the soft iron from N to S'; the number returning to S through the magnet is very small indeed.

84. Susceptibility. The magnetisation produced in a steel bar by a given magnetising force depends very greatly on the nature and temper of the steel, and can be largely modified by heating and annealing.

A bar of soft iron can be much more strongly magnetised by a given force than a bar of steel, the steel however retains much more of its magnetism than the iron unless indeed the latter be very specially treated. The ratio of the intensity of magnetisation produced by a given force to that force is known as the susceptibility, thus soft iron has a higher susceptibility than steel; the magnetism which remains after the magnetising force has been removed is spoken of as *permanent* or *residual* in contrast to the *temporary* magnetism which disappears with the force. Bodies which retain a large portion of their magnetism, as permanent or residual, are said to have a large "coercive force." In steel, then, the coercive force is high, and depends greatly on the temper of the steel.

For permanent magnets the steel is usually annealed to a blue temper, and may with advantage contain a small quantity of tungsten or molybdenum.

85. Influence of Temperature. When a steel magnet is heated to a moderate extent it loses some of its magnetism. Part of this it usually recovers on cooling—unless it has been raised to too high a temperature. If it be raised to a bright red heat and allowed to cool, when free from the action of magnetic force, it loses all its magnetism. In the case of moderate heating the permanent loss on cooling depends on the extent of the heating, and on the previous history of the magnet; if for example a magnet be "aged" by repeatedly heating it up to say 100° C. magnetising it as completely as possible when at that temperature, and allowing it to cool slowly, then after a long succession of such changes its permanent magnetism is not seriously affected by heating to a lower temperature than that reached in the "ageing" process. For a magnet so treated the magnetic moment decreases uniformly as the temperature rises, increasing again with a falling temperature to its original value; the relation between the magnetic moment and the temperature is expressed by the equation $M = M_0(1 - qt)$ where M_0 is the moment at the freezing point, M the moment at a temperature of $t°$, and q the coefficient of change of moment for each degree of temperature.

86. Maximum Strength of the Pole of a Magnet. The strength of the poles of an ordinary bar magnet having a length of say 10 centimetres or more, and a cross section of about 1 square centimetre, may be from 100 to 200 units; with a long thin bar the strength per unit area of the cross section of the bar may be much greater than this. The magnetic moment of a bar 10 centimetres long with a pole strength of 200 would be 10×200 or 2000.

Since the volume of such a bar would be 10 cubic centimetres the intensity of its magnetisation would be 200.

87. Theory of Magnetisation. We have already stated that if a magnet be broken into two pieces each part becomes a magnet; this is readily shewn by magnetising a thin knitting needle. On breaking it into two and dipping the ends of either half into iron filings, the filings adhere to both ends; if each of the two halves be now broken into two the portions into which they are divided have each two poles. This process can be continued as long as the pieces of steel are large enough to be broken. We are thus led to infer that magnetism may be a molecular phenomenon, and that the magnetic forces observed are the resultants of those due to the molecular magnets which make up the whole.

Thus, suppose we have a large number of similar, equal and equally magnetised small magnets. Let them be placed in a row as in Fig. 84, in such a way that the north pole of one is adjacent to the south pole of the next.

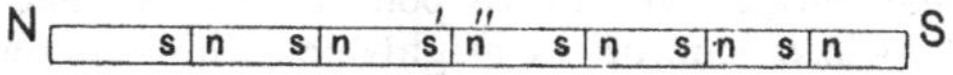

Fig. 84.

Then except at the two ends the magnetic repulsion of any north pole n'' will be neutralised by the attraction of the adjacent south pole s'; the resultant force due to the compound bar will consist of a repulsion from one end N combined with an attraction to the other end S; the bar will behave as a simple solenoidal magnet. We may thus picture to ourselves the effect of an actual magnet as due to the magnetism of its molecules.

Again, consider a test-tube full of iron filings; the tube is magnetic, but it is not a magnet. Bring it however into a magnetic field, so that its lower end, for example, is near the north pole of a strong magnet, and tap it gently; on removing it we shall find that the lower end has become a south pole. In the magnetic field the filings have set themselves so that their axes are along the lines of force, just as we saw them do when observing the lines of force on the glass plate, and the iron has thus become a magnet. On shaking the tube up again it ceases to be a magnet.

88. Molecular Magnets. In considering this molecular magnetism, the question arises whether the molecules

9—2

of iron are or are not permanent magnets. If they are
permanent magnets, the process of magnetisation will consist
in arranging them so that their axes all point in the same
direction; if they are not permanent magnets, the process
will consist firstly in magnetising each molecule, and then in
arranging the assemblage. If the first assumption be the
true one, and each molecule of a magnetic substance is itself
a magnet, we have to explain why every piece of iron is not a
magnet. This explanation was given by Weber, who pointed
out that the axes of the molecular magnets would set in all
directions, so that on the whole there would be no magnetic
force from a piece of iron, while Professor Ewing extended
this by pointing out that under the mutual forces between
the magnets any large assemblage of small magnets would set
themselves so as to form closed circuits, and would thus have
no external effect. The north pole of one magnet attracts
the south pole of one of its neighbours, while its own south
pole is attracted by some other north pole. The lines of force
thus form closed circuits within the magnets; few if any
escape beyond the assemblage.

Professor Ewing illustrated this by experimenting with a
large number of small compass-needles, resting on pivots,
arranged in regular order on a horizontal board; the compass-needles set themselves in the various manners shewn in
Fig. 85; if they are temporarily disturbed from these positions

Fig. 85.

they fall into some other having similar properties, and so
long as they are free from external force they as a rule exert
no magnetic force themselves. He was able to shew that
by varying the grouping of these small magnets many of the
magnetic properties of iron and steel can be very closely
imitated.

Thus suppose that by means of some impressed magnetic force the old grouping is broken up and a new one established; if this new grouping be very stable it will continue when the magnetic force is withdrawn : the assemblage resembles a steel magnet; if the stability of the new grouping be very small then on removing the magnetic force the magnets rearrange themselves on the slightest disturbance, so

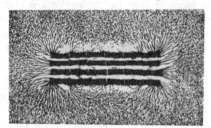

Fig. 86[1].

as to produce no external field, the assemblage is like a piece of soft iron. All these changes can go on under the mutual magnetic forces between the molecules ; the assumption of a force arising from friction or some similar action tending to hold them in their positions is not needed.

Fig. 87[1].

[1] These Figures are taken by the author's kind permission from Watson's *Text-book of Physics*. Longmans.

We may also examine the action of a number of small magnets by tracing by means of iron filings the lines of force they produce; if the magnets be arranged in order with all their like poles pointing the same way as in Fig. 86, there is a very distinct external field observable; if however they be allowed to take up the kind of positions shewn in Fig. 87, the external field is very small; we suppose then that in an unmagnetised piece of iron the molecules are arranged as in Fig. 87. When the iron is magnetised they are as in Fig. 86.

If the arrangement is stable so that it is retained after the magnetising force is removed, we have the case of a permanent magnet, if however it is unstable the iron is only temporarily magnetised.

89. Magnetic Force due to a Closed Cycle. The fact that a closed ring of magnetic particles exerts no external force is easily shewn by experiments with a piece of watch-spring. On magnetising it and dipping it into iron filings they adhere most strongly near the ends; along the length of the spring very few are visible. Now bend the spring so that its north polar end is brought into contact with the south polar end, and holding it in this position again cover it with iron filings. Very few adhere to it, the ring behaves almost as though it were unmagnetised.

CHAPTER X.

MAGNETIC CALCULATIONS.

90. Magnet in a Uniform Field. In a number of cases we can calculate the force between two magnets, or the force on one magnet due to a given field; the simplest is that in which the field is uniform.

Let us suppose that the strength of the uniform field is H and that it acts in the direction AB, Fig. 88. H is then the force with which a pole of unit strength would be urged in the direction from A to B. The force on a pole m would be mH, that on a pole $-m$ would be $-mH$, or mH acting from B to A.

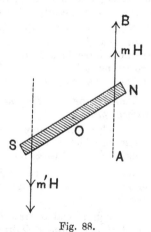

A magnet has two poles, let us suppose m is the strength of one pole, $-m'$ that of the other. Then the force on the one pole will be mH, that on the other $-m'H$, and the force on the magnet will be the resultant of these.

We shall see directly that m is equal to m'.

Fig. 88.

Now since we have two parallel forces acting on the magnet, the resultant consists of a force $mH - m'H$, or $(m - m')H$ in the direction of H, and a couple which tends to turn the magnet until its axis points in the direction of H.

We can shew however by experiment that in a case such as this the resultant is a couple only, the resultant force is zero; from this it follows that m is equal to m', or the strengths of the two poles of a magnet are equal.

The experiment is most easily performed when the earth's field is taken as the uniform field.

EXPERIMENT 19. *To shew that the earth's magnetic field is directive only, and that therefore the quantities of positive and negative magnetism in a magnet are equal.*

Place a bar magnet on a circular block of wood and allow it to float in a large vessel of water; bring the floating magnet to rest in some position in which its axis does not point north and south, and then gently remove the hands without disturbing the magnet. When set free it will be found that the block on which the magnet rests turns round, but that it does not acquire a movement of translation through the water; it follows from this that there is a couple on the magnet, but no resultant force.

Now if we call m the whole quantity of positive, m' the whole quantity of negative magnetism in the magnet, the resultant force in a field H is $mH - m'H$, and since this is zero we must have $m = m'$. That is the amounts of positive and negative magnetism in any magnet are equal.

PROPOSITION 5. *To find the couple on a magnet in a uniform field.*

Consider first a simple magnet, Fig. 89, having a north pole N, at which m units of magnetism are concentrated, and a south pole S with $-m$ units.

Let O be the centre of the magnet, $2l$ its length, and let the axis SN make an angle θ with AOB, the direction of the force H.

Through N and S draw LN and SK parallel to the direction of H, and through O draw KOL perpendicular to AOB.

Then the forces we have to consider are mH through N, acting along LN, and mH through S along KS.

Taking moments about O we have for the moment of the resultant couple $mH \times OL + mH \times OK$.

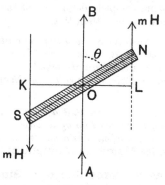

Fig. 89.

Now $\qquad OL = ON \sin \theta = l \sin \theta,$

and $\qquad OK = OS \sin \theta = l \sin \theta.$

Hence the moment required

$$= 2mlH \sin \theta = MH \sin \theta,$$

if M is the moment of the simple magnet.

Now we may look upon any magnet as compounded of a series of such simple magnets. The resultant action on each of these will be a couple, hence the whole resultant is a couple which we may write as before $MH \sin \theta$, where M is the moment of the magnet, and θ the angle which its axis makes with the direction of H.

It is clear that the couple is greatest when θ is 90°, for then $\sin \theta = 1$. In this case the axis of the magnet is at right angles to the direction of H.

The couple is zero when the axis coincides with the direction of H, for then $\sin \theta$ is zero.

These results are obvious.

Again, if H is unity and θ is 90°, the value of the couple is M.

We thus arrive at the result that the magnetic moment of a magnet is measured by the maximum couple it can experience when placed in a field of unit strength, and we thus obtain two somewhat more complete definitions of the terms magnetic axis and magnetic moment than those already given.

DEFINITION. *Suspend a magnet by its centre of gravity in a uniform magnetic field. Then it will be found that there is a line through the centre of gravity which always sets in a fixed direction in space. This line is called the* **Magnetic Axis** *of the magnet.*

DEFINITION. *It is found that a magnet placed in a uniform field is in general acted on by a couple. The ratio of the moment of this couple when it is a maximum to the strength of the field is called the* **Magnetic Moment of the Magnet.**

91. Magnetic Force due to a Simple Magnet.
Let NOS, Fig. 90, be a simple magnet and P any point near at which the magnetic force is required.

Let $OP = r$, $\angle PON = \theta$.

Let m and $-m$ be the pole strengths and $2l$ the length of the magnet. Then the force at P is the resultant of the forces due to m at N and $-m$ at S.

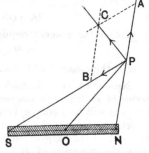

The force due to N is m/PN^2 acting along NP, that due to S is m/SP^2 acting along PS.

Produce NP and in it take

Fig. 90.

PA equal to m/PN^2; in PS take PB equal to m/SP^2 and complete the parallelogram $APBC$. Then the diagonal PC represents the resultant force.

We can find a mathematical expression for this resultant in terms of r, θ and the known quantities m and l, but this becomes complicated except in some special cases, *e.g.*, when P is on the axis of the magnet so that θ is zero or when OP is at right angles to the axis so that PON is $90°$.

PROPOSITION 6. *To find the force* due to *a simple magnet at a point on its axis produced.*

Let NS, Fig. 91, be the magnet, O its centre and P the point on ON produced at which the force is to be found.

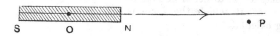

Fig. 91.

Let $$OP = r.$$

Then the two forces due to N and S act in the same straight line but in opposite directions and the resultant R is their difference.

Hence
$$R = \frac{m}{NP^2} - \frac{m}{SP^2}$$

$$= \frac{m}{(r-l)^2} - \frac{m}{(r+l)^2}$$

$$= \frac{m\{(r+l)^2 - (r-l)^2\}}{(r-l)^2 (r+l)^2}$$

$$= \frac{4mrl}{(r^2 - l^2)^2} = \frac{2Mr}{(r^2 - l^2)^2}.$$

We may put the expression into the form

$$R = \frac{2M}{r^3 \left(1 - \dfrac{2l^2}{r^2} + \dfrac{l^4}{r^4}\right)}.$$

In many cases l is so small compared with r that $(l/r)^4$ may certainly be neglected while it may also happen that to the accuracy with which we are working $2l^2$ may be neglected compared with r^2, *e.g.* if $l = 2$ cm. and $r = 20$ cm. then $2l^2/r^2 = \cdot 02$, and if we do not mind an error of 2 per cent. we may treat l^2 as zero compared with r^2.

If we suppose l to be so small that we may put $(l/r)^2$ zero in the expression for the force, we have

$$R = \frac{2M}{r^3}.$$

Thus we see that in this case the force at a given point due to a magnet is inversely proportional to the cube of the distance of the point from the centre of the magnet.

This position of the magnet is sometimes known as the "end-on" position; the axis of the magnet is directed end-on to the point at which the force is being calculated.

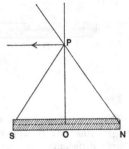

PROPOSITION 7. *To find the force due to a simple magnet at a point such that the line joining it to the centre of the magnet is perpendicular to its axis.*

Let NOS (Fig. 92) be the magnet, P the point at which the force is required, OP being perpendicular to NS.

Fig. 92.

Then since $NP = PS$ the two forces due to m and $-m$ are equal, each being m/NP^2. The resultant force bisects the angle between NP produced and PS, it is parallel therefore to NS.

Resolving the forces parallel to NS we have then for the resultant

$$R = \frac{m}{NP^2} \cos PNO + \frac{m}{SP^2} \cos PSO,$$

also

$$NP = SP,$$

$$\cos PNO = \cos PSO = ON/NP.$$

Thus

$$R = \frac{2mON}{PN^3} .$$

But

$$ON = l, \quad PN^2 = r^2 + l^2.$$

Hence

$$R = \frac{2ml}{\left(r^2 + l^2\right)^{\frac{3}{2}}} = \frac{M}{\left(r^2 + l^2\right)^{\frac{3}{2}}} ,$$

and if we again take a case in which l^2 is negligible compared with r^2 we have

$$R = \frac{M}{r^3} .$$

Thus in this position also—which is known as the broad-side-on position—the force on a magnetic pole is inversely proportional to the cube of its distance from the centre of the magnet.

But since we have

$$R = \frac{2M}{r^3} \text{ end-on position,}$$

and $$R = \frac{M}{r^3} \text{ broadside-on position,}$$

we see that, at equal distances from the centre, the force in the end-on position is twice as great as in the broadside-on position.

The expressions we have just found represent the magnetic intensities or forces on a unit pole at the two points; if a pole of strength m' be placed at either point, the forces acting on this pole will be respectively

$$2Mm'/r^3 \text{ and } Mm'/r^3.$$

92. Forces on one Magnet due to a Second. To find the force on a second simple magnet we must re-member that it has two poles; it will be necessary to calculate the force on each pole and to find their resultant, and the process is complex. When however the second magnet is

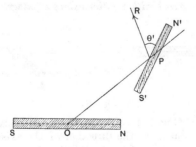

Fig. 93.

small and at some distance from the first we may simplify the calculation greatly; for though the field due to the first magnet is not strictly uniform, if N', S' (Fig. 93), the poles of the second magnet, are near together and at some distance from N and S, we may as a first approximation treat the force at N' as equal to that at S', each being equal to that at P the centre of the second magnet.

Hence we have to deal with the case of a small magnet in a uniform field, and if R be the strength of this field, θ' the angle between the direction of R and the axis of the second magnet, and M' the moment of this magnet, the resultant action on $N'S'$ is a couple $M'R \sin \theta'$ tending to decrease θ'.

In the end-on position the direction of R is along OP, θ' is the angle between PN' and OP produced and

$$\text{moment of couple} = \frac{2MM' \sin \theta'}{r^3}.$$

In the broadside-on position R is at right angles to OP, θ' is the angle between PN' and a line through P parallel to NS,

$$\text{moment of couple} = \frac{MM' \sin \theta'}{r^3}.$$

These two expressions are so important that we will find them in another way.

PROPOSITION 8. *To find the moment of the couple acting on a small magnet placed with its centre at a point on the axis of a second small magnet, assuming the distance between the centres to be large compared with the dimensions of either.*

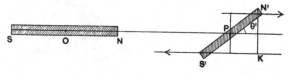

Fig. 94.

Let $N'PS'$ (Fig. 94) be the second magnet placed with its centre at a point P in the direction of the axis SON of the first magnet.

Let $OP = r$ and $OPS' = \theta'$, let m' be the pole strength and $2l'$ the length of the second magnet and M' its magnetic moment.

Then if $N'S'$ is small compared with OP the forces due to NS at N', S' and P may be taken as equal, and each is equal to $2M/r^3$.

Thus we have to find the resultant of two equal and opposite parallel forces $2m'M/r^3$ at N' and S' respectively.

Draw $S'K$ parallel to NP and $N'K$ to meet $S'K$ in K at right angles to $S'K$.

The resultant of the two parallel forces is a couple whose moment is

$$\frac{2m'M}{r^3} \times N'K.$$

Now $N'K = N'S' \sin N'S'K = 2l' \sin \theta'$.

Thus moment of couple required

$$= \frac{4m'l'M \sin \theta'}{r^3}$$

$$= \frac{2MM' \sin \theta'}{r^3}.$$

PROPOSITION 9.   *To find the moment of the couple acting on a small magnet placed with its centre so that the line joining it to the centre of a second magnet is at right angles to the axis of that magnet, the distance between the centres being large compared with the dimensions of either magnet.*

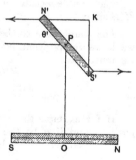

Here again the forces at P, N' and S' (Fig. 95) may be treated as equal, and equal to M/r^3.

This force acts at right angles to PO on a pole of strength m'.

Let $N'K$ be perpendicular and $S'K$ parallel to PO.

Fig. 95.

Then the resultant required is a couple whose moment is

$$m' \cdot \frac{M}{r^3} \times S'K.$$

Substituting for $S'K$ its value $2l' \sin \theta'$ we have

$$\text{moment of resultant couple}^1 = \frac{MM' \sin \theta'}{r^3}.$$

PROPOSITION 10. *A magnet hanging freely is deflected from its equilibrium position in which its axis is north and south by a distant magnet. To find the position it will take up.*

Let the magnet $S'N'$ (Fig. 96) come to rest with its axis making an angle ϕ with the north and south line, and let H be the horizontal component of the force due to the earth which acts on it.

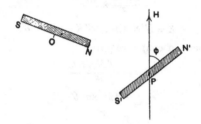

Fig. 96.

The couple acting on the magnet due to the earth is $M'H \sin \phi$.

The couple due to the distant magnet can be calculated if we know its position, let it be $M'G$. Then for equilibrium we must have

$$M'G = M'H \sin \phi,$$

or $$\sin \phi = \frac{\text{couple due to distant magnet}}{H}.$$

If for example the distant magnet be placed in the end-on position and if θ' be the angle between its axis and that of the deflected magnet we have

$$M'G = \frac{2MM' \sin \theta'}{r^3}.$$

Hence $$H \sin \phi = \frac{2M \sin \theta'}{r^3}.$$

---

[1] It should be noted that these expressions do not include the effect of the earth on the magnet.

Two cases of interest occur in practice :

(i) When the axis of the distant magnet points east and west (Fig. 97). In this case PO is at right angles to the direction of H, and $\theta' + \phi$ is a right angle, and

$$\sin \theta' = \cos \phi.$$

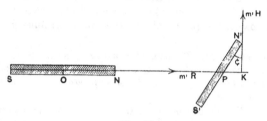

Fig. 97.

Hence $$H \sin \phi = \frac{2M \cos \phi}{r^3}.$$

Whence $$\tan \phi = \frac{2M}{Hr^3}.$$

(ii) When OP is at right angles to the deflected position of the suspended magnet as in Fig. 98. In this case $\theta' = 90°$ and $\sin \theta' = 1$.

Thus

$$\sin \phi = \frac{2M}{Hr^3}.$$

These two positions are spoken of as the tangent and sine positions respectively.

If the distant magnet be used in the broadside-on position similar results can be

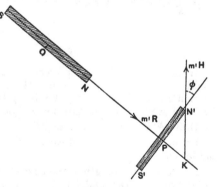

Fig. 98.

obtained but the couple due to it will be $MM' \sin \theta'/r^3$ instead of $2MM' \sin \theta'/r^3$, and hence we shall obtain

$$\tan \phi_1 = \frac{M}{Hr^3}$$

and
$$\sin \phi_1 = \frac{M}{Hr^3}$$

for the two cases, ϕ_1 being the deflexion.

If we call H the controlling and R the deflecting field we notice that the tangent law holds when the deflecting field is at right angles to the controlling, while the sine law holds when the deflecting field is perpendicular to the magnet.

We may if we please obtain the above formulæ directly thus.

PROPOSITION 11. *The axis of a small magnet points east and west. A second small magnet pivoted at its centre is placed with its centre at a point on the axis produced. To find the position it takes up.*

Let SON (Fig. 99) be the first magnet, P the point on the axis produced at which the centre of the second magnet $S'PN'$ is.

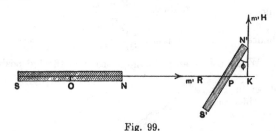

Fig. 99.

If it were not for the presence of NS the magnet $N'S'$ would point north and south. It is deflected from this position by the force due to NS, let ϕ be the angle between PN' and the undisturbed position PK of the second magnet.

Draw $N'K$ perpendicular to OP.

Let m' be the strength of each pole of $N'S'$, M' its moment, and let $OP = r$.

Let H be the strength of the field at N' before the magnet NS is brought near, and let R be the strength of the field due to NS. If OP is large compared with $N'S'$ the values of R at N', S' and P are equal. Hence since $OP = r$,

$$R = \frac{2M}{r^3}.$$

Now N' is acted on by two magnetic forces $m'H$ parallel to KN' and $m'R$ parallel to PK while S' is acted on by equal forces in opposite directions. Since the magnet which can turn about P is in equilibrium the resultant of each pair of forces must pass through P, and the forces are proportional to the sides of the triangle $N'KP$ to which they are parallel.

Hence $$\frac{m'R}{m'H} = \frac{KP}{N'K} = \tan\phi,$$

or $$\tan\phi = \frac{R}{H} = \frac{2M}{Hr^3}.$$

Thus the tangent law is proved.

PROPOSITION 12. *A small magnet pivoted at its centre is deflected by a second magnet. The axes of the two magnets are always at right angles, and that of the deflecting magnet passes through the centre of the first magnet. To find the position it takes up.*

Let SON (Fig. 100) be the first magnet, $N'PS'$ the second. Then P lies on ON produced, and $N'S'$ is at right angles to OP.

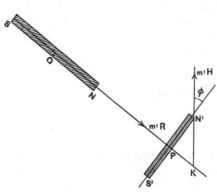

Fig. 100.

Let R be the strength of the field due to NS at P, since $N'S'$ is small, R is the strength at N' and S', and its direction is parallel to NP.

Let H be the strength of the undisturbed field at P, and let KN' be its direction, and let the angle $PN'K$ be ϕ.

Then the forces on N' are $m'H$ and $m'R$, while S' is acted on by equal forces in opposite directions, and for equilibrium we must have the forces proportional to the sides of the triangle $N'KP$, to which they are parallel.

Thus
$$\frac{m'R}{m'H} = \frac{KP}{N'K} = \sin \phi.$$

Also
$$R = \frac{2M}{r^3}.$$

Hence
$$\sin \phi = \frac{2M}{Hr^3}.$$

The formulæ for the broadside-on position are proved similarly; the figures will differ, and the value of R is M/r^3.

A more general case is proved in the same way.

Let a magnet $N'S'$ (Fig. 101) hanging in a field of strength H be deflected through an angle ϕ from its equilibrium position by a field R, and let the direction of R make an angle ψ with the original field.

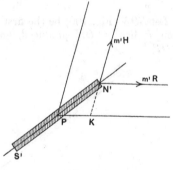

Thus in the figure
$$PN'K = \phi, \quad N'KP = \pi - \psi.$$

Hence $N'PK = \psi - \phi$.

Therefore
$$\frac{m'R}{m'H} = \frac{PK}{N'K} = \frac{\sin \phi}{\sin (\psi - \phi)},$$

or $\quad R = H \dfrac{\sin \phi}{\sin (\psi - \phi)}.$

Fig. 101.

93. Sine and Tangent Laws. We shall require to use these sine and tangent formulæ frequently. It is desirable that the student should realise that they depend merely on mechanical principles, and do not, except in the expression for R in terms of M and r, involve any magnetic theory.

Consider a rod ACB (Fig. 102) pivoted at its centre so that it can turn in a vertical plane, and let two strings be attached to it at A. Let one of these strings carry a

weight W. If the other string be left free, the rod will hang
in a vertical position under the pull, due to the weight W,
and we will call this the controlling force. Now apply a force

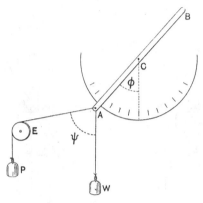

Fig. 102.

P to the second string. The rod will be deflected by this
deflecting force, and if a graduated circle be attached to a
board behind the rod with its centre at C, the angle of
deflexion can be measured. This angle of deflexion which we
will call ϕ, will depend on the magnitude and direction of P,
and we can find a relation between P, W, ϕ and ψ, the angle
between P and W, which will be exactly the same as that
given by the equations of the preceding Section.

Let the string AE, by which the force P is applied, pass
over a pulley E, and carry a small scale-pan into which
weights can be put. The force P will be measured by these
weights, including of course that of the scale-pan. By
making the distance AE large it can be arranged that AE
is practically horizontal for all positions of the magnet, but in
a more complete form of the apparatus the position of E can
be adjusted to secure this. By varying the position of E, the
direction of the force can be varied. Prof. Ayrton has devised
several pieces of apparatus for readily doing this; one simple
plan due to him is shewn in Fig. 103.

EXPERIMENT 20.   *To verify the tangent law.*

In this case it is necessary that the deflecting force should
be at right angles to the controlling force, *i.e.* AE must be
horizontal.

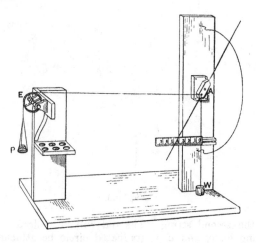

Fig. 103.

Let the weight W be 500 grams, and place 100 grams in
the scale-pan. Adjust the pulley until the string AE is
horizontal. This can be secured by observing when it is
parallel to a horizontal line marked on the apparatus, and
when this is the case observe the angle of deflexion ϕ, by
means of the pointer attached to the rod. Now change P,
making it say 200 grams, and again adjusting the pulley
observe a new value for ϕ.

Proceeding in this way, make a table of the corresponding
values of P/W and $\tan\phi$. It will be found that the two
quantities are equal.

Thus when the deflecting force is at right angles to the
controlling force, the ratio of the two is equal to the tangent
of the angle of deflexion.

EXPERIMENT 21. *To verify the sine law.*

In this case it is necessary that the deflecting force should be perpendicular to the rod.

Proceed as in the last experiment, but adjust the position of E until the line EA is at right angles to the pointer, determining when this is the case by aid of a set-square, and thus obtain a series of values of P and ϕ.

In this case we shall find that the values of P/W and $\sin \phi$ are equal.

Thus when the deflecting force is perpendicular to the deflected bar, the ratio of the deflecting to the controlling force is equal to the sine of the angle of deflexion.

The magnetic formulæ just proved can be submitted to a direct experimental verification by the aid of a simple piece of apparatus.

A small magnet is mounted on a pivot in a box with a glass cover, and carries a long light pointer, the position of which can be read on a horizontal circular scale with its centre at the pivot. The length of the pointer is usually perpendicular to the magnet. The scale is mounted on a long

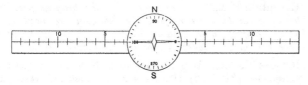

Fig. 104.

narrow board (Fig. 104). This rests on levelling-screws, and a V-groove is cut in the board in such a way as to pass through the centre of the circle. A straight scale of centimetres, the zero of which coincides with the pivot, is fixed parallel to the groove, and the position of the centre of a small magnet placed in the groove can be read off on the scale.

It is convenient that the graduations on the circle should be so arranged that when the axis of the magnet is at right angles to the groove, the pointer reads zero on the circle. Such an apparatus constitutes a simple form of magnetometer. It can be used in various ways; thus we can employ it to verify the tangent formula. This as we have seen can be written

$$\tan \phi = \frac{2M}{Hr^3},$$

where M is the moment of the deflecting magnet, H the strength of the controlling field.

We may put this result in the form

$$\tfrac{1}{2} r^3 \tan \phi = \frac{M}{H}.$$

Now M/H is constant so long as the moment of the deflecting magnet and the strength of the controlling field are unchanged; if then we measure r and ϕ, and the formula is true, this value of $\tfrac{1}{2} r^3 \tan \phi$ ought to be constant.

The sine formula can be written

$$\tfrac{1}{2} r^3 \sin \phi = \frac{M}{H},$$

which shews that under the proper conditions $\tfrac{1}{2} r^3 \sin \phi$ is constant.

EXPERIMENT 22. *To prove that in the tangent position $\tfrac{1}{2} r^3 \tan \phi$ is constant, where ϕ is the deflexion produced by a deflecting magnet in the end-on position with its centre at a distance r from the deflected magnet or compass-needle.*

Remove the deflecting magnet to a distance. Place the magnetometer (Fig. 105) with its V-groove east and west. If

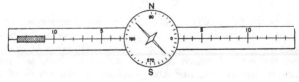

Fig. 105.

the pointer has been adjusted to be at right angles to the axis of the magnet, this is done by setting the pointer parallel to the scale.

Read the position of the pointer; if the circle has been adjusted as described in the last experiment, the pointer will read zero.

Place the deflecting magnet in the groove at some convenient distance from the deflected magnet, and read on the scales the distance between the centres r and the angle of deflexion ϕ_1.

Reverse the deflecting magnet, replacing it in the groove with its centre in the same position as previously, but with its north pole where its south pole was.

The needle will be deflected, but in a direction opposite to that of its previous motion. Read the deflexion ϕ_2.

If everything be perfectly adjusted, ϕ_1 will be equal to ϕ_2; if the two do not differ greatly, the mean $\frac{1}{2}(\phi_1 + \phi_2)$, which we will call ϕ, will be free from errors due to the fact that the pointer is not accurately at right angles to the deflected magnet, and that the magnetic centre of the deflected magnet may not be midway between its ends.

If great accuracy is wanted remove the deflecting magnet from the groove and replace it on the other side of the needle, but at the same nominal distance from it as previously, and read the deflexions ϕ_1' and ϕ_2' as before. The mean of the four $\frac{1}{4}(\phi_1 + \phi_2 + \phi_1' + \phi_2')$ will eliminate errors which may arise from the centre of the needle not being exactly over the centre of the scale.

Now remove the deflecting magnet to another distance r_2, and repeat the observations. Do this for five or six distances, and then make a table, containing in consecutive columns the values of r, ϕ_1, ϕ_2, $\phi\left[=\frac{1}{2}(\phi_1 + \phi_2)\right]$, r^3, $\tan\phi$ and $\frac{1}{2}r^3\tan\phi$.

It will be found that the numbers in the last column are approximately constant.

We thus verify the result that for a given deflecting magnet and control field the quantity $\frac{1}{2}r^3\tan\phi$ is constant, and the theory shews us that this constant is the ratio of the moment of the magnet to the strength of the field.

It may easily happen that while the deflected magnet is so small that our approximate formulæ will hold for it, the length of the deflecting magnet can not be treated as very small compared with the distance r.

In this case we must have recourse to a more complete formula.

The fundamental result $R/H = \tan\phi$ holds, but the formula for R is less simple than we have assumed.

Let $2l$ be the distance between the poles of the deflecting magnet—$2l$ will really be less than the length of the magnet, and cannot be determined with any accuracy, but if we treat the magnet as solenoidal, we may use for $2l$ the length of the magnet, and we shall then introduce rather too large a correction.

Then we have seen, Section 91, that the value for R is

$$R = \frac{2Mr}{(r^2 - l^2)^2}.$$

Hence instead of the formula

$$\frac{M}{H} = \tfrac{1}{2} r^3 \tan\phi,$$

we have

$$\frac{M}{H} = \frac{1}{2}\frac{(r^2 - l^2)^2}{r}\tan\phi,$$

and if we are working to a degree of accuracy which does not permit of neglecting the ratio l^2/r^2, we should verify that $\tfrac{1}{2}(r^2 - l^2)^2 \tan\phi/r$ is constant.

We observe that the value of the constant is M/H, so that if we know M we can use the observation to find H or conversely if we know H we can find M.

EXPERIMENT 23. *To prove that in the sine position the ratio $\tfrac{1}{2} r^3 \sin\phi$ is constant, where ϕ is the angle of deflexion produced by a deflecting magnet with its centre at a distance r from a deflected magnet or small compass-needle.*

We can use the magnetometer for this experiment also. Place the deflecting magnet in position, and then turn the whole instrument (Fig. 106) round until the pointer on the circle reads zero, *i.e.* is parallel to the groove. The groove of course no longer points east and west. Now remove the deflecting magnet to a distance. The needle will move and come to rest with its axis north and south. Read the pointer. The reading ϕ_1 will be the deflexion of the needle produced by the deflecting magnet in the given position.

Reverse the deflecting magnet and proceed as above to find ϕ_2. Then form a table giving values of r, ϕ_1, ϕ_2, $\phi \equiv \tfrac{1}{2}(\phi_1 + \phi_2)$, r^3, $\sin\phi$ and $\tfrac{1}{2} r^3 \sin\phi$.

The last series of numbers will be approximately constant,

and if the same deflecting magnet is used as in the tangent law experiments, the value of the constant M/H will be the same as for those experiments.

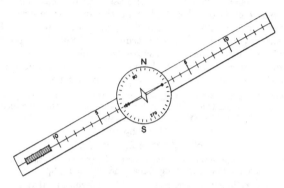

Fig. 106.

If $2l$ the length of the deflecting magnet is too long to be neglected, we have to replace, as in Experiment 23, the quantity r^3 by $(r^2 - l^2)^2/r$.

The formulæ obtained for the broadside-on position may be verified in the same way, but the apparatus is not quite so convenient for this, for the deflecting magnet has to be placed with its length at right angles to, instead of parallel to, the groove, and the groove will run north and south.

94. Law of the Inverse Square. The relation between the deflexions in the end-on and broadside-on positions requires further consideration, for if ϕ and ϕ' be the two deflexions respectively observed with the distance r, the same in the two cases, then

$$\tfrac{1}{2} r^3 \tan \phi = \frac{M}{H} = r^3 \tan \phi'.$$

Hence $\qquad\qquad \tan \phi' = \tfrac{1}{2} \tan \phi.$

This can be verified by observing the deflexion ϕ in the end-on position, then placing the deflecting magnet in the broadside-on position, and observing the value of ϕ' for the same value of r.

The results we have just arrived at afford a conclusive proof of the Law of the Inverse Square. The results that $\frac{1}{2}r^3 \tan\phi$ or $\frac{1}{2}r^3 \sin\phi$ are constant, and that $\tan\phi/\tan\phi' = 2$ have both been arrived at on the assumption that the law of force between two poles is that of the inverse square. Experiment proves these results true, and we infer therefore that the inverse square law is true also. Now it is possible to arrange apparatus to measure the deflexions with much greater accuracy than with the magnetometer just described, when this is done the law is very fully verified.

Gauss, the great German magnetician, shewed that if we suppose the law of force between two poles to be $1/r^n$ instead of $1/r^2$, then the accurate tangent formulæ for the two positions become

$$\tan\phi = L_1 r^{-(n+1)} + L_3 r^{-(n+3)} + \ldots$$

and

$$\tan\phi' = L_1' r^{-(n+1)} + L_3' r^{-(n+3)} + \ldots,$$

where L_1, L_1' are numerical coefficients depending on the magnetic moment, and on the controlling force, and where $L_1/L_1' = n$.

Now Gauss found as the result of his experiments that he could express $\tan\phi$ and $\tan\phi'$ by the two series

$$\tan\phi = 0\cdot086870r^{-3} - 0\cdot002185r^{-5}$$

and

$$\tan\phi' = 0\cdot043435r^{-3} + 0\cdot002449r^{-5},$$

and these held for values of r from about 1 to 4 metres.

These formulæ contain a double verification of the law. In the first place the value of $n+1$ is 3, or n is equal to 2.

In the second the ratio of L_1/L_1', which by theory is n, is

$$\cdot086870/\cdot0043435,$$

and this also is exactly 2.

Thus we may infer from these experiments that to a very high degree of accuracy the force between two magnetic poles is inversely proportional to the square of the distance between them.

CHAPTER XI.

MAGNETIC MEASUREMENTS.

95. Experiments with the Magnetometer. We can use the magnetometer described in the previous sections for various other experiments.

EXPERIMENT 24. *To compare the magnetic moments of two magnets, or of the same magnet after various treatments.*

(*a*) By the method of equal distances.

Adjust the magnetometer so that the groove points east and west; and observe the reading of the pointer. Place one of the two magnets on the scale, with its north end pointing to the needle, and at some convenient distance away and observe the deflexion, let it be ϕ_1. Remove this magnet to a distance, and place the second magnet on the scale taking care that its centre is in exactly the same position as that of the first. Let the deflexion be ϕ_2. Then if M_1, M_2 are the moments of the two magnets, we have

$$M_1 = \tfrac{1}{2} H r^3 \tan \phi_1,$$
$$M_2 = \tfrac{1}{2} H r^3 \tan \phi_2;$$

and hence $\qquad M_1 : M_2 = \tan \phi_1 : \tan \phi_2.$

Thus the magnetic moments are compared.

The value to be chosen for r—the "convenient distance"—depends on the length and strength of the magnet; it should be as large as possible, consistent with giving a deflexion which can be measured with sufficient accuracy. Suppose for example that the circle is such that we can read the position of the needle to a quarter of a degree, and that we want to know M to 1 per cent. Then we have to find for what value of ϕ an error of $0°·25$ makes an error which is less than ·01 in $\tan \phi$.

A table of tangents will shew that this is the case if ϕ is greater than $25°$ and less than $65°$. Thus the distance should be such that the deflexion is not less than $25°$. So far as the measurement of $\tan \phi$ is concerned, greater accuracy will be secured by reducing the distance so as to make ϕ about $45°$, but when the formula depends on the assumption that $2l^2/r^2$ may be neglected, and if r is made too small this condition can not be fulfilled. It may of course happen that if the magnet is weak the value of r required to give a deflexion of even $25°$ is too small to permit of the neglect of $2l^2/r^2$ and in this case the required accuracy can not be attained ; we must have recourse to some more delicate method of reading the deflexion.

If the two magnets happen to be of the same length, or nearly of the same length, it is not necessary that $2l^2/r^2$ should be so small as to be negligible.

For if l_1, l_2 be the lengths we have

$$M_1 = \tfrac{1}{2} \frac{H (r^2 - l_1{}^2)^2}{r} \tan \phi_1,$$

$$M_2 = \tfrac{1}{2} \frac{H (r^2 - l_2{}^2)^2}{r} \tan \phi_2.$$

Hence
$$\frac{M_1}{M_2} = \frac{(r^2 - l_1{}^2)^2}{(r^2 - l_2{}^2)^2} \frac{\tan \phi_1}{\tan \phi_2},$$

and if l_1 is equal to l_2 this ratio is $\tan \phi_1/\tan \phi_2$ for any value of r. If $(l/r)^4$ can be neglected while $(l/r)^2$ must be retained we have

$$\frac{M_1}{M_2} = \left\{ 1 - \frac{2 (l_1{}^2 - l_2{}^2)}{r^2} \right\} \frac{\tan \phi_1}{\tan \phi_2},$$

and the formula is still true if $2 (l_1{}^2 - l_2{}^2)/r^2$ is sufficiently small.

It is clear that the accuracy of the result is increased by taking the four readings of ϕ as described in Experiment 22, *i.e.* by placing the centre of the magnet first to the east then to the west of the needle and in each of these positions observing (i) with the north pole pointing east, (ii) with the north pole pointing west.

(b) By the method of equal deflexions.

Place the first magnet in the end-on position at a distance r_1 from the needle so as to give a convenient deflexion ϕ. Remove the first magnet, and place the second with its centre at a distance r_2, adjusting this until the deflexion is again ϕ.

Then
$$M_1 = \tfrac{1}{2} H r_1{}^3 \tan \phi,$$
$$M_2 = \tfrac{1}{2} H r_2{}^3 \tan \phi.$$

Hence
$$M_1 : M_2 = r_1{}^3 : r_2{}^3.$$

The remarks made under (a) as to the choice of values of r and ϕ apply here. The experiment may be modified by placing the first magnet in position at a distance r_1 so as to give a deflexion ϕ, and then placing the second magnet on the opposite side of the needle at a distance r_2 so as to bring the pointer back to zero.

The forces on the needle due to the two magnets are thus equal and opposite, one produces a deflexion ϕ, the other a deflexion $-\phi$. Hence we have

$$\frac{2M_1}{r_1^3} = \text{force due to first magnet} = \frac{2M_2}{r_2^3}.$$

Hence $M_1 : M_2 = r_1^3 : r_2^3.$

EXPERIMENT 25. *To measure the magnetic moment of a magnet having given a magnetometer in a field of known strength.*

Place the magnet with its axis east and west, and in the end-on position at a known distance r from the needle of the magnetometer, and observe the deflexion ϕ.

Then, Proposition 10, Section 92, $M = \frac{1}{2}r^3 H \tan\phi,$

and if H is known, M can be found. As before greater accuracy is secured by taking observations with the magnet in the four positions.

If H be the horizontal component of the earth's field, its value in England is about 18 in C.G.S. units. See Section 102.

Examples. (1) *A magnet placed with its centre at a distance of 20 cm. from a magnetometer needle deflects it 35°. Find its moment assuming the value of H to be ·18 units.*

Hence $M = \dfrac{8000}{2} \times \cdot 18 \times \tan 35$

$$= 4000 \times \cdot 18 \times \cdot 700$$

$$= 504.$$

(2) *The length of the magnet is 4 cm., find the strength of either pole assuming them to be at the ends, and obtain a more correct expression for the magnetic moment, allowing for the fact that l^2/r^2 is not very small.*

For the strength of the poles we have

$$\text{Strength} = \frac{\text{Magnetic Moment}}{\text{Length}} = \frac{504}{4} = 126 \text{ units.}$$

Or in other words each pole of the magnet if it were isolated would exert a force of 126 dynes on a unit pole at a distance of 1 cm. or of 1·26 dynes on a unit pole at a distance of 10 cm.

To correct the value of the magnetic moment already found for the length of the magnet, we have in this case $2l = 4$, $l = 2$.

$$M = \tfrac{1}{2}H \frac{(r^2 - l^2)^2}{r} \tan \phi$$

$$= ·09 \times \frac{(400 - 4)^2}{20} \times ·700$$

$$= ·09 \times 7840·8 \times ·700$$

$$= 494,$$

and the strength of each pole is 123·5. Thus the error made by neglecting $(l^2/r)^2$ in this case is 2 per cent.

96. The Mirror Magnetometer.

More accurate results can be obtained in many of these experiments by the use of a mirror magnetometer. If we have to use a pointer made of aluminium, or of a fibre of glass, or some other such material, the graduated circle on which the deflexions are read cannot well be more than 10 to 20 cm. in diameter, and besides difficulties are introduced by the friction at the pivot which carries the magnet.

In the mirror magnetometer (Fig. 107) the magnet is attached to the back of a small mirror which is suspended from a suitable support by a fine silk or quartz fibre; rays of light from a slit in front of a lamp fall on the mirror, and are reflected on to a scale, placed at right angles to the line joining the slit and mirror, and form there an image of the slit. As the mirror moves, this image moves on the scale, and since the distance between the mirror and scale may be considerable, say from one to two metres, a very small angular motion of the mirror produces a considerable motion of the spot on the scale. The pointer attached to the magnet is in this case virtually

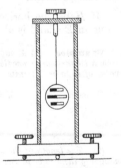

Fig. 107.

the beam of light and as this may easily be from 20 to 100 times as long as any possible material pointer, the sensitiveness is greatly increased.

The mirror used in such a magnetometer may either be plane or concave; if a plane mirror is used a convex lens of suitable focal length is necessary in order to form the real image. The lens, which is shewn at L in Fig. 108, is placed between

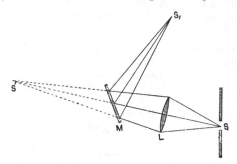

Fig. 108.

the slit S and the mirror M in such a position that S' the real image of the slit which would be formed by the lens if the mirror were removed would be as far behind the mirror, as the scale is in front of it. The light reflected from the mirror passes by the lens and forms at S_1 a reflected image of S'. This is a real image of the slit.

Sometimes the lens is placed close to the mirror so that the reflected as well as the incident light passes through it. In this case, Fig. 109, S is at the principal focus of the lens. The rays from S after traversing the lens fall as a parallel pencil on the mirror, and are reflected as a parallel pencil; this is brought to a focus at S_1 on the scale, which is approximately at the same distance from the lens as the slit S.

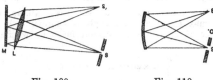

Fig. 109.   Fig. 110.

If a concave mirror is used the slit is placed at the same

distance from the mirror as its centre of curvature, Fig. 110;
the rays falling on it are reflected to S_1, forming a real image
at the same distance from the mirror as the slit.

It remains to consider how the displacement of the spot on
the scale is connected with the angular motion of the mirror
and magnet.

Let S be the slit, Fig. 111, and S_1 its reflected image,
suppose that in the undisturbed condi-
tion the adjustments are such that S
and S_1 coincide, or rather that S_1 is just
vertically above S. When the magnet
is disturbed let MN be the direction of
the normal to the mirror; then originally
N coincided with S, and NMS is the
angle through which the magnet has

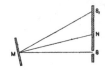

Fig. 111.

been deflected, let this · be ϕ; let SS_1 the displacement on
the scale be a centimetres, and let SM be d centimetres.
Then since the ray SM is reflected along MS_1 we have

$$\angle SMN = \angle NMS_1.$$

Hence
$$\angle SMS_1 = \angle 2SMN = 2\phi,$$

or the spot of light is deflected through twice the angle
through which the magnet is turned.

Moreover, since S_1SM is a right angle, we have

$$S_1S = SM \tan SMS_1.$$

Hence
$$a = d \tan 2\phi.$$

If then we know d and observe a, the displacement of the spot,
we can find the angle 2ϕ from a Table of Tangents, and
hence we can obtain ϕ the deflexion of the magnet.

In practice when this method is used ϕ is small, and we know that
the circular measure of a small angle is approximately equal to its
tangent. Hence if ϕ is required in degrees, instead of writing $a/d = \tan 2\phi$
we may write

$$\frac{a}{d} = 2 \times \text{circular measure of } \phi.$$

Hence
$$\phi = \frac{1}{2} \frac{a}{d} \frac{180°}{\pi}.$$

However, we usually require to know not ϕ but $\sin \phi$ or $\tan \phi$ and in the case where ϕ is small we have as an approximate result, sufficient for our purpose,

$$\sin \phi = \tan \phi = \frac{1}{2} \tan 2\phi = \frac{1}{2} \frac{a}{d}.$$

Example. *The slit is at a distance of 105 cm. from the mirror, and the displacement is 5·5 cm. Find the angle through which the mirror is deflected.*

Here
$$\tan 2\phi = \frac{5 \cdot 5}{105} = \frac{1 \cdot 1}{21} = \cdot 0524.$$

Thus
$$2\phi = 3° \text{ and } \phi = 1° 30'.$$

According to the approximate formula

$$\phi = \frac{\cdot 0524}{2} \times \frac{180}{\pi} = 1° \cdot 502$$

$$= 1° 30' 12'' \cdot 6.$$

Thus the difference is $12'' \cdot 6$ which is so small that for our purposes it may be neglected.

In some other arrangements a horizontal scale is placed before the mirror and a telescope is adjusted so as to view the image of the scale reflected from the mirror; as the magnet moves, the image of the scale seen in the telescope moves also.

The experiments already described can be repeated with this more delicate apparatus.

97. Measurement of the Strength of a Uniform Field and of a Magnetic Moment. We have already seen how to determine the quantity M/H, the ratio of a magnetic moment of a magnet to the strength of the field in which it is placed so that if we know the value of the moment we can find the strength of the field and conversely.

We shall now see how we can find the value of MH the product of the magnetic moment and the strength of the field and hence if both MH and M/H are known we can find M and H.

If we take a magnet and suspend it by a fine fibre so that its axis hangs in a horizontal position in a field of strength H, it will oscillate about its position of equilibrium in which, if the field be due to the earth, its axis would point north and south.

If the axis be displaced through a small angle θ from its equilibrium position the couple tending to bring the magnet back will depend on $M.H$ being equal to $M.H \sin\theta$. If we can find this couple experimentally we can obtain MH. Now we know that if θ is small, $\sin\theta$ is very approximately equal to θ so that the couple may be written $MH.\theta$, thus the couple is proportional to θ the displacement from rest. In a case such as this the magnet oscillates backwards and forwards and it can be shewn, both by theory and experiment, that the time of an oscillation is a constant[1]. The time will depend on the shape and mass of the magnet, being greater if the magnet is big and heavy, than it is if the magnet is light. It will also depend on the restoring couple being less when this is big, than when it is small.

We can shew from some dynamical reasoning that the time of swing T is given by the formula

$$T = 2\pi \sqrt{\frac{K}{MH}},$$

where K is a quantity called the moment of inertia of the magnet and depends on its form and mass.

From this we find

$$MH = \frac{4\pi^2 K}{T^2}.$$

If then we can calculate K and determine T by experiment, this equation gives us the value of MH.

Now K can be found by measurement. If the magnet be a circular cylinder of mass m grammes, length $2l$ and radius a centimetres,

$$K = m\left(\frac{l^2}{3} + \frac{a^2}{4}\right).$$

If it be rectangular, $2l$ being the length, $2a$ the breadth in a horizontal direction, then

$$K = m\left(\frac{l^2 + a^2}{3}\right).$$

---

[1] Glazebrook's *Dynamics*, § 146.

EXPERIMENT 26. *Having given a magnet of known moment of inertia, to find the value of MH.*

Let K be the known moment of inertia. Then MH is given by the formula

$$MH = \frac{4\pi^2 K}{T^2},$$

where T is the time of a complete oscillation. To observe T suspend the magnet with its axis horizontal, and protect it with

a small bell jar or some other covering to shield it from draughts—a convenient arrangement is shewn in Fig. 112, in which the magnet is suspended inside a wide-mouthed bottle, the bottom of which has been removed—make a mark on the glass opposite one end of the magnet when at rest, or on a sheet of paper under the magnet, and set the magnet oscillating through a small angle by bringing a second magnet near, and then removing it.

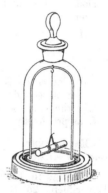

Fig. 112.

Determine by means of a stop watch the time occupied by a number of swings of the magnet. To do this start the stop watch as the end of the magnet passes the mark, and count the consecutive transits reckoning the first as 0, the second as 1 and so on. Allow the magnet to swing for some time and stop the watch just as it passes the mark when making the mth transit. Observe the number of seconds the watch has been going. By dividing this by the number of transits we get the time between two transits. By a complete period is meant the time between two transits in the same direction. Multiply therefore the observed time between two transits by 2 and we have T the time of a complete period. Substitute in the formula

$$MH = 4\pi^2/T^2, \text{ we get } MH.$$

The time during which the magnet is allowed to swing must depend on circumstances; if the magnet will go on swinging and we can count the number of transits without making a mistake, the longer it is the more accurate the result.

If the magnet has a period of 8 or 10 seconds it will usually be sufficient to observe some twenty transits corresponding to a total interval of from $1\frac{1}{2}$ to 2 minutes.

We can put this formula into another form which is sometimes more convenient.

Let n be the number of transits in 1 second. Then since a complete period is the interval between 2 transits in the same direction, $n/2$ will be the number of complete periods in 1 second, and $2/n$ will be the time of a complete period.

Hence
$$T = 2/n$$

and
$$MH = \frac{4\pi^2}{T^2} K = \pi^2 n^2 K.$$

To find n divide the number of transits observed by the number of seconds in which they have occurred. On substituting the value so found the same result will be obtained for MH as previously.

EXPERIMENT 27. *To determine the moment of a magnet and the strength of the field in which it hangs.*

Find as in Experiment 26 the value of MH and as in Experiment 25 the value of M/H. On solving the two equations thus obtained we get M and H.

The two equations are
$$MH = \pi^2 n^2 K,$$
$$\frac{M}{H} = \tfrac{1}{2} r^3 \tan \phi.$$

Multiplying them together we have
$$M^2 = \tfrac{1}{2}\pi^2 n^2 r^3 K \tan \phi,$$
and dividing the first by the second,
$$H^2 = \frac{2\pi^2 n^2 K}{r^3 \tan \phi}.$$

Hence both M and H can be found.

Example. The moment of inertia of a magnet is 380 c.g.s. units. When allowed to swing freely in a field of strength H twenty transits are observed in 2 minutes 19 seconds and when placed at a distance of 30 cm.

from a magnetometer needle the deflexion is $10° 30'$. Find the value of M and of H.

We have $$n = \frac{20}{139} = \cdot 144.$$

Thus $$MH = n^2\pi^2K = 77\cdot4,$$

$$\frac{M}{H} = \tfrac{1}{2}r^3 \tan\theta = 2498.$$

Whence $$M = 440 \text{ c.g.s. units,}$$
$$H = \cdot176 \text{ c.g.s. units.}$$

EXPERIMENT 28. *To compare the strengths of two magnetic fields.*

For this purpose we make use of the formula $MH = \pi^2n^2K$, for it is clear that if the same magnet be swung in different fields the number of oscillations in a second will vary, the field strength being proportional to the square of the number of oscillations per second.

Thus if n_1, n_2 be the number of complete oscillations per second, H_1, H_2 the field strengths in the two positions, we have since M and K are the same

$$MH_1 = \pi^2n_1{}^2K$$
$$MH_2 = \pi^2n_2{}^2K,$$

and hence $$H_1 : H_2 = n_1{}^2 : n_2{}^2.$$

We must remember in using this method that there is a magnetic field due to the earth; if all magnets be removed from the neighbourhood of the swinging magnet it will oscillate in the earth's field only, if another magnet be brought near the field will be the resultant of the earth's field and that due to the second magnet.

EXPERIMENT 29. *To compare the strength of the field at a point on the axis of a magnet produced with that due to the earth and to find hence the magnetic moment of the magnet.*

Allow the vibration magnet to oscillate under the earth's field alone and determine the time of twenty transits.

Find hence n_1 the number of oscillations in one second.

Place the bar magnet with its axis north and south in such a position that its south pole is to the north of the centre of the vibration magnet and points to it.

Let the field due to the bar magnet at the centre of the vibration magnet be F. The resultant field is $F + H$ and if in this case n_2 is the number of oscillations per second

$$\frac{F + H}{H} = \frac{n_2^2}{n_1^2}.$$

Hence

$$\frac{F}{H} = \frac{n_2^2 - n_1^2}{n_1^2}.$$

Thus F is found if H is known.

If the distance between the centres of the two magnets be r centimetres and if r is considerable compared with the length of the bar magnet, then

$$F = \frac{2M}{r^3},$$

where M is the moment of the bar magnet.

Hence

$$M = \tfrac{1}{2} r^3 H . \frac{n_2^2 - n_1^2}{n_1^2}.$$

If the bar magnet is very long compared with the distance x between its north pole and the centre of the vibration magnet, then if m is the strength of either pole of the bar magnet, we have approximately

$$F = \frac{m}{x^2}.$$

Hence

$$m = x^2 F = x^2 H \frac{n_2^2 - n_1^2}{n_1^2}.$$

98. Determination of the axis of a magnet.

If a magnet be suspended in the earth's field so that it can move about a vertical axis, it will set with its axis north and south.

The magnetic meridian is the vertical plane which contains the direction of the earth's force.

If the direction of the magnetic meridian at the place of observation be known we can from the above fact find the position of the axis of the magnet; it is the direction in the magnet which coincides with the meridian. Generally, however, the north and south line is not known with accuracy

and we proceed to shew how to determine both it and the
axis of the magnet.

Let O (Fig. 113) be the centre of the magnet, AOB its
axis, COC' a line, marked on one face of the magnet, passing
through O.

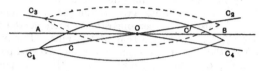

Fig. 113.

Assume for the present that the position of the north
and south line is known.

Lay the magnet down on a sheet of paper so that this
face is horizontal while AOB points north and south and make
marks on the paper opposite to the points C and C'. Let
them be C_1 and C_2.

Remove the magnet and turn it over so that the face
which was in contact with the paper is uppermost, and the
face on which the line CC' is drawn next to the paper.
Replace it so that AOB is again in the magnetic meridian.

Make marks on the paper under the new positions of
C and C'.

Let them be C_3, C_4.

Then since the axis of the magnet and the line CC' are
both fixed in the magnet, the angle between them is constant.
And since in one position CC' is over C_1C_2 and in the other
over C_3C_4 it is clear that the axis of the magnet and therefore the magnetic meridian bisects the angle between C_1C_2
and C_3C_4.

We can use this result to find both the axis of the magnet
and the magnetic meridian thus :

EXPERIMENT 30. *To determine the axis of a magnet and
to find the magnetic meridian at any point.*

The magnet is supported by a stirrup from which it can be easily withdrawn and replaced with the face which was uppermost turned downwards. The stirrup is suspended over a sheet of paper by a fine silk fibre from which the torsion has been carefully removed. Two marks C, C' are made one at each end of the magnet. Place the magnet in the stirrup in such a way that its axis is horizontal and allow it to come to rest. It will rest with its axis in the magnetic meridian. Stick a pin[1] into the paper opposite to each of the marks C, C', let C_1, C_2 be the position of the pins. Remove the magnet from the stirrup and replace it in the inverted position. Stick pins C_3, C_4 into the paper opposite the new positions of C and C'.

Join C_1, C_2 and C_3, C_4, and bisect the angle between these lines.

This bisector gives the magnetic meridian and the axis of the magnet is that direction in the magnet which is parallel to the meridian.

[1] The pins should be of brass.

CHAPTER XII.

TERRESTRIAL MAGNETISM.

99. Magnetism of the Earth. The magnetic force due to the earth varies from point to point on its surface both in direction and in amount. The direction of the lines of force is not in general horizontal but makes an angle with the horizontal plane through the point of observation which depends on the position of that point. If a piece of un-magnetised steel be suspended from its centre of gravity it will rest in any position in which it is placed; if it be magnetised it will set in a definite position and the north pointing end will in these latitudes point downwards. If the steel could be freely suspended accurately from its centre of gravity the direction of its axis would give the direction of the earth's field at the point; it is difficult to do this, and so two instruments are used, in one of these a magnet is supported so that it can move in a horizontal plane. The position of its axis when it comes to rest gives the magnetic north and south, and a vertical plane passing through this is called the plane of the magnetic meridian, the line in which this axis cuts the earth's surface is the direction of the meridian.

In the other instrument a magnet can turn about a horizontal axis through its centre of gravity. The instrument is set so that this axis is at right angles to the magnetic meridian, the magnet then moves in the plane of the meridian and the direction of its axis when it comes to rest gives the direction of the earth's force.

It is clear that the line of action of the earth's force lies in the plane of the meridian and that it can be resolved into two components, horizontal and vertical respectively, in this plane.

DEFINITION. *The angle between the plane of the magnetic meridian and the true north and south line—the astronomical meridian—is called the* Declination.

DEFINITION. *The angle between the direction of the resultant force and a horizontal line drawn in the plane of the magnetic meridian is called the* Dip *or* Inclination.

Let I be the intensity of the earth's field, i the dip and δ the declination.

Then the direction of I makes an angle i (Fig. 114) with the intersection of the magnetic meridian and a horizontal plane at the point of observation. We can resolve I into a horizontal component H and a vertical component V, and we have

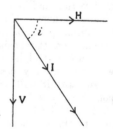

$$H = I \cos i,$$
$$V = I \sin i.$$
Thus $\qquad V = H \tan i,$
and $\qquad I = H \sec i.$

Hence if we can determine H and i we can calculate the vertical component and the total intensity.

Fig. 114.

100. Measurement of the Dip. We have already seen (Experiment 27) how to determine the strength of a horizontal magnetic field though of course additional refinements are introduced in accurate instruments.

To find the Dip we use a dip circle (Fig. 115). A light lozenge-shaped magnet can turn about a very fine horizontal axis which passes through its centre of gravity. This axis which in an accurate instrument rests on two polished agate knife-edges passes through the centre of a vertical graduated circle, so that the magnetic axis of the magnet forms a diameter of the circle. When disturbed the magnet moves parallel to the plane of the circle, and when at rest the position of its ends can be read off on the circle. For this purpose microscopes are attached to the instrument. The circle can turn about a vertical axis and can thus be set in the plane of the magnetic meridian.

To use the instrument the circle is turned round a vertical axis until the axis of the magnet is itself vertical.

When this is the case the plane of the circle, parallel to which the magnet moves, is at right angles to the meridian.

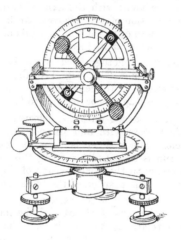

Fig. 115.

On turning the circle then through a right angle the magnet will swing in the plane of the meridian. Allow it to come to rest and read the position on the circle of either end of the magnet; the zero of the circle is adjusted to be in the horizontal plane. Thus if all the adjustments are complete the reading obtained gives the dip.

If the axis of the magnet does not pass through the centre of the circle an error will be introduced. This is eliminated, however, by determining the dip from both ends of the magnet and taking the mean.

Further precautions are needed to eliminate other possible sources of error, but these we cannot go into here.

101. Measurement of the Declination. An approximate method of doing this has already been described § 98. In the more delicate apparatus as used at Kew and elsewhere the magnet is hollow. At one end it carries a scale photographed on glass, at the other a lens whose focal length is equal to that of the magnet. Light from any point on the scale then emerges as a pencil of parallel rays from the lens. The scale is viewed through a telescope which can turn about a vertical axis coincident with the axis of suspension of the magnet, and the position of the telescope can be read off on a horizontal circle whose centre lies on the axis of rotation; the telescope has cross wires at its focus.

A point on the photographed scale is selected—this corresponds to the point C of Fig. 113 above and the line joining it to the centre of the lens which is known as the line of collimation of the magnet corresponds to the line CC'.

The telescope is turned until the selected point coincides with the vertical cross wire and its position read on the horizontal circle. The magnet is then dismounted and inverted, the stirrup is arranged for doing this readily and the telescope is moved until the same division of the scale is again on the cross wire. The position of the telescope is read.

The plane of the magnetic meridian bisects the angle between the two positions of the axis of the telescope, and is thus determined relatively to the circle. The plane of the geographical meridian can be found by observations on the sun, the instrument is usually arranged to make this possible, and the angle between these two planes is the declination.

From the observations of horizontal intensity and dip we can calculate the total intensity and the vertical intensity. If we also know the declination we can calculate the intensity in any given direction.

102. Magnetic Survey of the Earth; Magnetic Maps. The values of the magnetic elements obtained by experiment are found to vary from point to point of the earth's surface; moreover the values found at any one place are found to alter slowly with the time. The table drawn up by

Dr Chree gives the values of these quantities at some important places as found in the year 1901.

TABLE.

Mean Values, *for the years specified, of the Magnetic Elements at Observatories whose Publications are received at the National Physical Laboratory.*

Place	Latitude	Longitude	Year	Declination	Inclination	Horizontal Force. C.G.S. Units	Vertical Force. C.G.S. Units
Pawlowsk	59 41 N.	30 29 E.	1899	0 34.1 E.	70 38.8 N.	·16536	·47078
Katharinenburg	56 49 N.	60 38 E.	1899	9 59.6 E.	70 39.7 N.	·17795	·50706
Kasan	55 47 N.	49 8 E.	1897	7 54.8 E.	68 34.8 N.	·18616	·47454
Copenhagen	55 41 N.	12 34 E.	1900	10 12.2 W.	68 39.0 N.	·17513	·4480
Stonyhurst	53 51 N.	2 28 W.	1901	18 9.7 W.	68 45.7 N.	·17348	·44638
Hamburg	53 34 N.	10 3 E.	1900	11 18.1 W.	—	·18152	—
Wilhelmshaven	53 32 N.	8 9 E.	1900	12 27.7 W.	67 44.0 N.	·18095	·44193
Potsdam	52 23 N.	13 4 E.	1900	9 56.3 W.	66 33.7 N.	·18844	·43466
Irkutsk	52 16 N.	104 16 E.	1899	2 1.5 E.	70 13.7 N.	·20133	·56009
de Bilt (Utrecht)	52 5 N.	5 11 E.	1899	13 54.7 W.	—	·18502	—
Kew	51 28 N.	0 19 W.	1901	16 48.9 W.	67 9.5 N.	·18451	·43804
Greenwich	51 28 N.	0 0	1900	16 29.0 W.	67 8.5 N.	·18450	·43764
Uccle (Brussels)	50 48 N.	4 21 E.	1900	14 13.6 W.	66 9.8 N.	·18952	·42896
Falmouth	50 9 N.	5 5 W.	1900	18 29.1 W.	66 45.2 N.	·18689	·43507
Prague	50 5 N.	14 25 E.	1900	9 7.0 W.	—	·19947	—
St Helier (Jersey)	49 12 N.	2 5 W.	1901	16 56.5 W.	65 42.7 N.	—	—
Parc St Maur (Paris)	48 49 N.	2 29 E.	1898	14 53.8 W.	64 58.3 N.	·19676	·42140
Vienna	48 15 N.	16 21 E.	1898	8 24.1 W.	—	·20797	—
O'Gyalla (Pesth)	47 53 N.	18 12 E.	1901	7 23.4 W.	—	·21175	—
Odessa	46 26 N.	30 46 E.	1898	4 41.5 W.	62 30.5 N.	·22033	·42341
Pola	44 52 N.	15 51 E.	1900	9 25.3 W.	60 15.9 N.	·22202	·38871
Nice	43 43 N.	7 16 E.	1899	12 4.0 W.	60 11.7 N.	·22390	·39087
Agincourt (Toronto)	43 47 N.	79 18 W.	1899	5 27.8 W.	74 33.5 N.	·16503	·59744
			1900	5 28.8 W.	74 32.5 N.	·16512	·59709
Perpignan	42 42 N	2 53 E.	1898	13 47.0 W.	60 1.7 N.	·22386	·38818
Tiflis	41 43 N.	44 48 E.	1898	2 5.5 E.	55 50.6 N.	·25635	·37784
Capodimonte (Naples)	40 52 N.	14 15 E.	1900	9 10.2 W.	—	—	—
Madrid	40 25 N.	3 40 W.	1898	15 51.3 W.	—	—	—
			1899	15 48.4 W.	—	—	—
Coimbra	40 12 N.	8 25 W.	1900	17 20.1 W.	59 24.3 N.	·22768	·38506
			1899	17 16.1 W.	59 19.6 N.	·22805	·38449
			1901				
Lisbon	38 43 N.	9 9 W.	1900	17 18.0 W.	57 54.8 N.	·23516	·37484
Tokio	35 41 N.	139 45 E.	1897	4 29.9 W.	49 2.8 N.	·29816	·34356
Zi-ka-wei	31 12 N.	121 26 E.	1899	2 20.3 W.	45 47.6 N.	·32825	·33747
Havana	23 8 N.	82 25 W.	1900	3 7.8 E.	52 36.0 N.	·30948	·4048
Hong Kong	22 18 N.	114 10 E.	1900	0 18.5 E.	31 24.7 N.	·36728	·22430
Tacubaya	19 24 N.	99 12 E.	1895	7 45.6 E.	44 22.2 N.	·33428	·32764
Colaba (Bombay)	18 54 N.	72 49 E.	1898	0 28.6 E.	21 6.2 N.	·37445	·14451
			1899	0 25.4 E.	21 13.9 N.	·37448	·14549
Manila	14 35 N.	120 59 E.	1899	0 51.9 E.	16 19.9 N.	·37981	·11130
			1900	0 52.1 E.	16 16.0 N.	·38029	·11096
Batavia	6 11 S.	106 49 E.	1898	1 14.9 E.	29 47.4 S.	·36752	·21040
Dar-es-salem	6 49 S.	39 18 E.	1898	8 18.1 W.	36 56.8 S.	·28966	·21785
Mauritius	20 6 S.	57 33 E.	1899	9 32.9 W.	54 16.8 S.	·23854	·33171
Rio de Janeiro	22 55 S.	43 11 W.	1900	7 55.7 W.	13 17.0 S.	·2504	·0592
Melbourne	37 50 S.	144 58 E.	1898	8 20.1 E.	67 22.4 S.	·23364	·56050

The magnetic condition of a country is best indicated on a magnetic map. Fig. 116 gives such a map constructed for the year 1900 from the results of Räcker and Thorpe's *Survey of the British Isles*. The dip is 68° near Swansea, Monmouth, Northampton and Ely. Thus a line drawn through these points will be a line of equal dip. Such a line is known as an **Isoclinal Line**.

By finding another series of points such as these which have the same dip and joining them, a second isoclinal can be drawn and so on. Thus a whole series can be constructed and from such a map the value of the dip can be found.

In the same way we can draw a series of lines each of which passes through points at which the declination is the same. Such a line is called an **Isogonal Line**.

Thus along a line passing near Hull, Lincoln, Northampton, Oxford, Salisbury and Swanage, the declination in 1891 was 18°. This line then is an isogonal and a number of such isogonals can be drawn.

In the same way a series of points at which the horizontal component is constant can be found, and from this the **Lines of equal horizontal force** can be constructed. The three sets of lines are all shewn in the map.

In a similar manner lines can be constructed for the earth. The line at which the dip is zero, and the axis of the dipping needle horizontal, passes round the earth in a position approximately coincident with the equator; the dip is 90°, and the needle stands vertical at the two magnetic poles. The north magnetic pole is approximately in Latitude 70° 5′ N. and Longitude 96° 43′ W. The position of the south magnetic pole is not known.

In the case of the declination as we pass to the west from England across the Atlantic the westerly declination increases and then decreases again gradually until we come to near the longitude of Lake Superior when we cross a line along which the declination is zero. This is called an **Agonic Line**; to the west of this line the declination is easterly, the magnetic needle points to the east of true north. If we travel to the

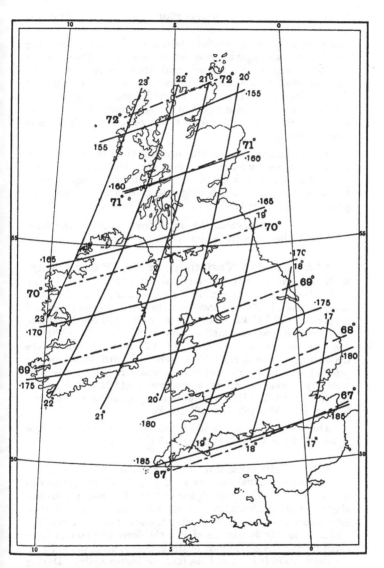

Fig. 116.

east from England the westerly declination decreases, becoming
zero along an agonic line which traverses Russia from
St Petersburg to Sebastopol, and then passes down the
Arabian Gulf to the west of India. To the east of this line
the declination is easterly. There is another agonic line
forming an oval enclosing part of China, Japan, and the north-
eastern part of Siberia. Within this oval the declination is
again westerly.

The horizontal force again increases from the value ·18 in
England as we travel south across the Atlantic, reaching a
maximum of about ·3 rather to the south of the equator and
falling again as the south magnetic pole is approached. In
parts of India and Cochin China it reaches the value ·38.

As we travel towards the north from England the force
falls. At the magnetic poles its value is zero.

103. Secular Variations of the Earth's Magnetism.

As we have already said, the values for the dip,
declination and force change with time. Thus in 1576 the dip in
London was 71° 50'. It increased up to 1720 when it reached
a maximum of 74° 42'. Since that time it has been decreasing
and its value at present at Kew is 67° 9'. In 1580 the
compass in London pointed 11° 17' east of north, the
declination was easterly. This decreased until 1657 when it
pointed true north, the declination was zero; it then became
westerly increasing up to a maximum of 24° 30' which it
reached in 1816. It is now decreasing again and at present
has at Kew the value of 16° 48' W.

The value of the horizontal intensity is increasing; at Kew
it was ·1716 in 1814, it is now ·1845 c.g.s. units.

104. Daily and Annual Variations.

In addition
to the above gradual changes a very slight daily change can
be observed. In the morning the westerly declination increases
slightly and continues to increase till about 1 p.m., it then
decreases somewhat rapidly during the afternoon and evening,
and more slowly during the night; the decrease becomes more
rapid during the early hours of the morning, until about 7 a.m.
the declination is least and the increase begins again. During

the summer the amount of this change is about 8′. In winter the change is less. There is also an annual change which seems to be related to the position of the sun in its orbit.

105. Magnetic Storms. In addition to these regular changes sudden disturbances of the magnetic elements occur from time to time, and these are often of considerable magnitude. At magnetic observatories instruments are installed for recording the changes which take place photographically.

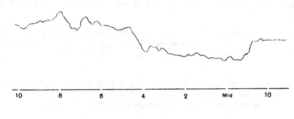

Fig. 117 a.

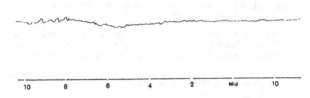

Fig. 117 b.

Fig. 117 a is a reproduction of such a curve from the declination instrument at the Kew Observatory, shewing a storm which occurred on April 10th, 1902. The trace for an ordinary, quiet day is also reproduced in Fig. 117 b.

EXAMPLES ON MAGNETISM.

1. Several soft iron needles are floating vertically very close to one another on small separate bits of cork in a basin of water. A powerful magnetic pole is held above the group. Describe and explain the movements that will take place.

2. What force does a magnetic pole of strength 6 units exert upon a pole whose strength is 16 units placed at a distance of 4 cm. away?

3. A pole of strength 8 units acts with a force of 4 dynes upon another pole placed at a distance of 6 cm. Find the strength of the latter pole.

4. A magnetic needle of pole strength 5 units and length 10 cm. is placed in a magnetic field of strength 12 so as to be at right angles to the lines of force. With what couple does the field act upon the needle?

5. A needle of magnetic moment 12 is placed in a magnetic field of strength 12 units in such a direction that its axis makes an angle of 30° with the lines of force. Find the couple acting on the magnet.

6. Three bar magnets, A, B and C, have the same intensity of magnetisation. A is 10 cm. long and 1 sq. cm. in section, B is 10 cm. long and 2 sq. cm. in section and C is 20 cm. long and 1 sq. cm. in section. Compare their magnetic moments.

7. Calculate the magnetic force at a point on the axis of a bar magnet 100 cm. distant from the centre of the magnet, the strength of each pole being 100 units and the length of the magnet being 4 cm.

8. A magnet whose pole strength is 2000 and length 20 cm. is placed on a table. Find the field produced at a point abreast of its middle point and 10 cm. distant from it.

9. The centres of two small magnets coincide and their axes are at right angles, the magnetic moment of the one being twice that of the other. Shew that the lines of force due to the combination are, at all points on the axis of the second magnet produced, inclined at 45° to that axis.

10. The magnetic moment of a small magnet is 36 c.g.s. units. Find the magnetic force due to it at a point in its axis produced, distant 18 cm. from its centre.

11. A magnet, suspended horizontally, is caused to oscillate at two different places. At the first place it makes 100 oscillations in 4 minutes, at the second 110 oscillations in 4 minutes. Compare the values of the horizontal components of the magnetic force at the two places.

12. What force or forces must be applied to a magnet, whose magnetic moment is M, to hold it fixed in an East and West position?

$(H = \cdot 18$ c. g. s. unit.$)$

13. A magnet is suspended by a wire so as to rest horizontally in the magnetic meridian. When the upper end of the wire is twisted through 90° the magnet is deflected 30° from the meridian. How much further must the upper end be turned to deflect the magnet 90° from the meridian?

14. A magnetic needle points North and South. A bar magnet pointing East and West is placed with its centre 50 cm. East of the needle, and it is found that the needle then points North-East. Find the magnetic moment of the bar magnet given that $H = \cdot 18$.

15. A magnet turning about a vertical axis makes 50 vibrations per minute at a place where the dip is 45°, while it makes 60 vibrations per minute at a place where the dip is 30°. Compare the resultant magnetic forces at the two places.

16. A needle makes 15 oscillations per minute in a certain magnetic field. How many will it make when re-magnetised so that its magnetic moment is half as great again as before?

17. A small magnetic needle, when swinging in the earth's magnetic field only, makes 8 oscillations per minute. A long magnet is placed with one of its poles at a distance of 8 cm. from the centre of the suspended needle, and in such a direction that the lines of force due to the magnet have, in the neighbourhood of the small needle, the same direction as those due to the earth. In this position the needle oscillates 12 times per minute. If the long magnet be now moved parallel to itself until its nearest pole is now at a distance of 12 cm. from the centre of the needle, calculate the rate at which the latter will now oscillate.

18. A magnet is placed with its axis on the magnetic meridian and its South pole pointing North. It is found that there is a neutral point at a distance of 14 cm. from the South pole of the magnet. If the length of the magnet be 10 cm. and $H = \cdot 18$ c. g. s. unit, find the strength of the poles of the magnet.

19. The maximum intensity of permanent magnetisation of a steel bar 10 cm. long and 1 sq. cm. in section has been found to be 225 c.g.s. units. Find the tangent of the greatest angle of deflexion of a magnetometer needle which such a magnet could cause if the needle be 30 cm. from the centre of the magnet.

$(H = \cdot 18$ c.g.s. unit.$)$

CHAPTER XIII.

THE ELECTRIC CURRENT.

106. Electric Currents. If two insulated bodies at
different potentials be connected to the opposite quadrants
of an electrometer the needle is deflected to an amount depending on the difference of potential. If the two bodies be
connected by a conductor this difference of potential disappears. A charge of positive electricity passes along the
conductor from the body at high to that at low potential of
just sufficient amount to equalize the potentials. This transference of the charge constitutes an electric current in the
conductor. In such a case the potentials are equalized with
great rapidity, the current is of very brief duration. It is
however possible by various means to maintain a steady
difference of potential between the conductors, even when
connected; and in this case the current in the wire is a continuous one, we can examine and measure its effects.

107. The Voltaic Cell. A voltaic cell is perhaps
the readiest means by which this potential difference can be
maintained. Such a cell in its simplest form consists of a
plate of zinc and a plate of copper, which dip separately into a
vessel containing dilute sulphuric acid. The copper is called
the positive plate of the cell, the zinc is the negative plate.
If the copper and zinc be connected by copper wires to the
opposite quadrants of an electrometer the needle shews a
difference of potential. This difference of potential is called
the electromotive force of the cell. Let us denote it by
E. The unit in which electromotive force is measured is

called a volt, after Volta, the discoverer of the cell. We shall consider later how to define this unit and how it is to be measured; meanwhile we must remember that when we say the electromotive force of a battery is E volts we mean that this difference of potential exists between the copper and the zinc plates, and hence that E volts measures the number of units of work required to carry a unit of positive electricity from the copper to the zinc plate. In this case when the plates are not connected together the cell is said to be on open circuit.

Two main theories, the contact theory and the chemical theory respectively, have been developed to account for this action, these we shall consider at a later stage.

Now connect the plates by means of a conductor. The potential difference indicated by the electrometer falls somewhat—the amount of fall is dependent on the nature of the conductor—but a potential difference is maintained and so a continuous current must flow in the conductor.

We may illustrate the process by considering two reservoirs filled with water to different levels. On opening communication between the two the water flows from the reservoir at higher level to that at lower until the two levels are equalized, when the current stops. If however water is being simultaneously pumped back from the lower to the upper reservoir a continuous current will be maintained

Or again, imagine two vessels (Fig. 118), A, B, connected by two pipes CD, EF; EF being at a higher level than CD. Let there be a tap G in EF and a turbine or water-wheel W in CD, which when it is turned causes water to flow from B to A.

On working the wheel the level is raised in the one vessel, lowered in the other; if a constant force be applied to the wheel this will go on until the pressure due to the difference of level balances that due to the wheel; the flow ceases, the difference of level being thus maintained at a steady height. The wheel is analogous to the battery, the vessels A and B correspond to the two conductors at different levels.

Now open the tap G. There is a flow along EF from A to B, the pressure in A tends to fall, that in B to rise, hence the turbine is now able to propel water from B to A along DC and a steady current is maintained. The levels in the two vessels are not the same as they were when G was closed, that in A having sunk to A', that in B risen to B'; since the flow is

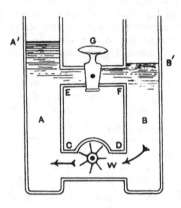

Fig. 118.

from A' to B' the level of A' is above that of B'; the difference between A' and B' corresponds to the potential difference between the conductors when in electrical connexion; while the difference before the tap was opened gives its value on open circuit.

There are various other forms of battery besides the simple voltaic cell and various other methods of producing a current; we shall recur to these later; as a matter of fact the simple cell described would not for various secondary causes give a steady current, in our experiments we may use a Daniell cell, § 127, or preferably a storage battery, § 131. In all cases however the electromotive force of the battery E volts measures the potential difference between its plates when on open circuit.

If the copper and zinc plates of a Daniell cell be connected by a wire the current in the wire is from the copper to the zinc.

108. Measure of a Current. Let us suppose that we have a wire the ends of which are connected with a battery whose electromotive force is E volts; a current is flowing in the wire and we must proceed to consider its measurement and its effects.

We measure a uniform current of water or other liquid flowing in a tube by the quantity of liquid which crosses any given section of the tube in the unit of time. In the same way a uniform current of electricity is measured by the number of units of electricity which cross any section of the conductor in 1 second.

A point of some importance should be noted here. Let AB (Fig. 119) be a tube of variable section through which water is flowing, the tube being full, and let P, Q be two sections of the tube; if the tube remains full, since the water is incompressible, the quantity of water which crosses P in any given time is equal to that which crosses Q in the same time; if the tube were full of air this would not necessarily be the case, the compression of the air between P and Q might vary and in consequence there might be more air between P and Q at one period of the flow than at another, if this were the case the current at P would not be always equal to that at Q.

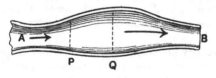

Fig. 119.

In this respect the flow of electricity when a steady condition has been reached resembles that of water. Careful experiment shews that the current at P is always equal to that at Q. Hence the current in a conductor is measured by the quantity crossing *any* section of the conductor per second.

109. Relation between Current and Quantity transferred. Now let the current be c. This means that in one second c units of electricity cross any given section of the conductor. Hence in t seconds the quantity transferred is ct units. Thus if we denote the quantity transferred by q we have

$$q = ct.$$

We may of course write this

$$c = \frac{q}{t}.$$

Thus to measure the current, assuming it uniform, we have to measure the number of units of electricity transferred in time t seconds and divide it by the time; the quotient gives the current.

We have already defined the unit quantity of electricity as measured electrostatically. If the quantity q be measured in these units, the current given by the ratio q/t will be in electrostatic units, and the electrostatic unit of current is that current in which an electrostatic unit of electricity is transferred across each section of the conductor per second. We shall find however that for many purposes this is not the most convenient unit to employ, when dealing with electric currents a much larger unit of current is chosen. This is called an **ampere** and will be defined later. It is sufficient to say here that experiment shews that a current of one ampere conveys about 3×10^9 electrostatic units of electricity per second across each section of the conductor in which it is flowing.

The quantity of electricity conveyed by 1 ampere flowing for 1 second will be the electro-magnetic unit of quantity. This is known as a coulomb.

DEFINITION. *A* **coulomb** *is the quantity of electricity conveyed by* 1 **ampere** *flowing for* 1 *second.*

We may compare this with the different units of length adopted for different purposes; in some cases it is convenient to measure in millimetres or even in thousandths or millionths of a millimetre, in others a kilometre is selected.

110. Tubes of Force and Electric Currents.

If we have two insulated conductors such as the plates of a condenser, the one of which is charged positively while the other is negative, lines of force pass as we have seen from the positive to the negative conductor, and the electric forces can be represented as arising from a tension along the lines of force combined with a pressure at right angles to them, each line of force starting from a unit positive charge terminates in a unit negative charge. In Fig. 120 the distribution of the lines of force due to a charged condenser AB is shewn. Now let the plates A, B be connected by a conducting wire CD. The tubes of force in the space occupied by the wire cannot exist within the material of the conductor. They shrink up into the wire, the ends which were on the condenser plates A, B travelling along the wire until they meet, and the effect of the tube is annulled; the pressure in the medium is thus relieved and the tubes in the neighbourhood of the wire close on to it: the unbalanced pressure in the surrounding

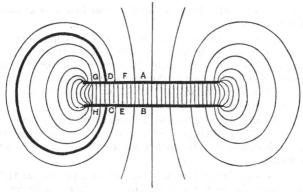

Fig. 120.

space forces other tubes up to the wire and these in their turn shrink up into it until all the tubes originally existing between A and B have passed into the wire and the field is annulled. From this point of view we may look upon the transient current in the wire as a transference of tubes of force across the field up to the wire within which they disappear. Now however suppose that A and B are connected to the plates of a battery. The battery by its action generates tubes of force as fast as they disappear in the wire, and the continuous current consists in the passage of these tubes across the field, their ends sliding as it were along the conductor until they are absorbed into the wire. Viewed in this aspect the current is made up of the transference of positive electricity in one direction combined with the equal transference of negative electricity in the other.

111. Effects due to an Electric Current. When a current passes through a conductor various effects shew themselves. These may be classified as magnetic, thermal, and chemical.

112. Magnetic Action of a Current. Magnetic force is exerted in the neighbourhood of a wire which carries a current. This was discovered by Oersted, a Danish professor, in 1820.

EXPERIMENT 31. *To shew that a current in a wire produces magnetic force.*

(a) Connect a wire to the two poles of a Daniell or other cell and hold it in a horizontal position above and parallel to a magnet pivoted at its centre in such a manner that the current from the copper to the zinc pole flows from south to north in the wire. It will be found that the north end of the magnet is deflected towards the west.

If the wire be held under the magnet the deflexion is to the east. If the direction of the current is reversed the magnet is deflected to the east when the wire is above, and to the west when it is below.

If the wire be held in an east and west position at right angles, that is, to the axis of the magnet, no deflexion is observed.

(β) Fix the wire in a vertical position and bring a small compass-needle near it. Notice that the compass always tends to set itself at right angles to the line drawn from its centre perpendicularly on to the wire.

(γ) Fix the wire so that it may pass at right angles through a sheet of stiff paper or cardboard supported in a horizontal position, and sprinkle iron filings on the card; on allowing the current to pass and tapping the card it will be found that the iron filings set themselves in concentric circles with their centres at the point in which the card is cut by the wire.

Thus it follows from these observations that there are lines of magnetic force round the wire, while the last two observations shew that these lines are circles in planes perpendicular to the wire. The wire moreover passes through the centres of these circles.

The direction of the force can also be determined from the first two observations, and various rules have been framed to express the law found.

Thus extend the right arm in a horizontal position with the palm downwards and the thumb pointing to the left. Imagine now a current to be running down the arm from the shoulder to the fingers and that the thumb-nail represents a north magnetic pole. Twist the arm round so that the thumb moves upwards at first. The direction of motion of the thumb gives the direction of the magnetic force due to the current in the arm.

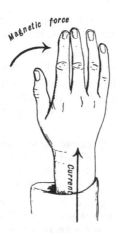

Fig. 121 *a*.

If the magnetic pole be above the wire the thumb must be held uppermost and the hand twisted in the same direction as before, the motion of the thumb still gives the direction of the force.

Or again we may state the rule thus :

Consider a right-handed screw—an ordinary wood screw—which is being screwed into a piece of wood. If a current flow along the screw from the head to the point, in the direction, that is, in which the point of the screw is moving, a north pole will tend to move round the current in the direction in which the screw is being turned.

If the current be reversed so that it moves from the point to the head, imagine the screw as being withdrawn from the wood ; the direction in which it is turned will still give the direction of the magnetic force.

Thus we may state :

If a right-handed screw be placed so that the direction of the current in a wire coincides with the direction of translation of the point of the screw when the screw is turned, a north magnetic pole near the wire will tend to move in the direction of rotation of the screw.

This relation between the direction of the current and that of the force is illustrated in Figs. 121 *a*, *b*.

Further experiments may be made to shew that a current exerts magnetic force.

Thus wind a piece of insulated wire into a long spiral and place a steel knitting-needle in the coil with its length along the axis of the spiral.

On passing a strong current through the wire and attempting to withdraw the steel it will be found that it is pulled into the coil and that it has become a magnet.

If the steel be replaced by a piece of soft iron, when the current passes the iron becomes temporarily a very powerful magnet, but loses much of its magnetism again when the current ceases to flow.

Fig. 121 *b*.

Such a magnet is known as an electro-magnet, and the

magnets used in dynamo machines and electric motors are of this class.

We shall see later how to use the magnetic effect of a current to measure the current, at present our object is to get a qualitative knowledge of the properties of a current.

113. Thermal Effects of a Current.   A current heats any conductor through which it passes.

The action of an incandescent lamp is an obvious illustration of this fact; the carbon filament is brought to a white heat by the current. Electric heaters of various forms in which this effect is utilized are not uncommon.

By the help of a calorimeter we can measure the heating effect produced by a current in a spiral of insulated wire or in a lamp. Thus if we have a water calorimeter[1] we immerse the spiral in a known mass m of water and observe the temperature of the water. On passing the current the temperature rises and the product of the total rise of temperature and the mass of the water gives, apart from losses from the surface of the calorimeter and other minor corrections, the amount of heat produced in the wire.

Other thermal effects are due to the passage of a current, *e.g.* the junction of two metals across which the current flows changes in temperature as the current passes, and the change depends on the direction of the current, if it passes in one direction the junction is heated, if the direction of the current is reversed the junction is cooled. A converse fact to this is that a current can be produced by heating the junction of two dissimilar metals. These effects however are usually small compared with the heating of the conductor.

114. Chemical Action of a Current.   If a current be allowed to pass through dilute acid or through an aqueous solution of a metallic salt, it is found that the liquid is decomposed by the current. This phenomenon is called electrolysis, and the liquids are known as electrolytes. Many fused

---

[1] Glazebrook, *Heat*, §§ 38—52.

salts are electrolytes, so are some solids, *e.g.* iodine of silver, and also probably certain gases.

In dealing with electrolysis certain terms introduced by Faraday will be found useful.

When a current traverses a liquid conductor the surfaces at which it enters and leaves the conductor are called *electrodes*, the surface at which it enters the conductor is the *anode*, that at which it leaves the conductor is the *kathode*.

Suppose now that we have two platinum plates immersed in dilute sulphuric acid and that a current is passed from one plate *A* through the liquid to the second plate *B*. Then *A* is the anode and *B* the kathode. Bubbles of gas collect on the two plates, and if the products be examined it will be found that the gas on the anode is oxygen while that on the kathode is hydrogen. Moreover it can be shewn that the volume of the hydrogen collected in any time is twice that of the oxygen. This is most easily done by the use of the apparatus shewn in Fig. 122 known as a water voltameter. The electrodes are two pieces of platinum foil with which contact can be made from the outside by means of platinum wire sealed through the glass. The vessel is filled with slightly acidulated water, the taps *C* and *D* being open; when the water has risen above the levels of the taps they are closed. On passing a current from *A* to *B* the gas from the anode rises in the inverted burette *AC*, that from the kathode in the other burette *BD*.

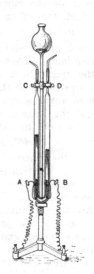

Fig. 122.

The two gases are thus kept separate, and it will be noticed that there is no apparent decomposition in the liquid between the electrodes. The graduations of the burettes serve to measure the volumes of the gases given off, and it will be found that, when allowance is made for the difference of pressure to which the two are subjected in consequence of the difference of level of the water surfaces in the two burettes, and for the difference in the solubility of the two gases, the volume of the hydrogen is twice that of the oxygen.

115. Observations on Electrolysis.

EXPERIMENT 32. *To illustrate the phenomena of electrolysis.*

In Fig. 123 AB represents a water voltameter, E, F, G, are three beakers, E and F contain a slightly acid solution of copper sulphate, while G is filled with silver nitrate.

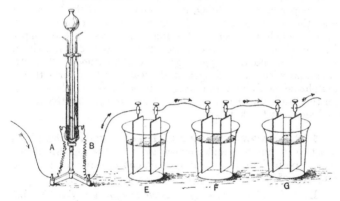

Fig. 123.

Two copper plates are placed in E, and two platinum plates in both F and G. These are connected up so that a current can pass through the water voltameter and through the liquids in the three beakers in series. Make the connexions and allow the current to pass for some time, say 15 minutes, then examine the results. If the electromotive force of the battery used has been sufficient (the reason for this proviso will appear later) oxygen has been given off at A, hydrogen at B. The copper anode in E is probably black and scaly— the extent of this depends in a great measure on the purity and exact condition of the materials—the copper kathode in E is covered with a bright coating of freshly deposited copper; the platinum anodes in F and G are unchanged, but oxygen gas has been given off from their surfaces during the experiment; the kathodes however are covered with copper and silver respectively. Electrolysis has gone on in all four vessels.

The products of the electrolysis are known as ions. The ion which appears at the kathode is the kation, the ion which appears at the anode is the anion. The metal of the solution is in each case the kation, and appears at the kathode; in this respect, as in some others, hydrogen behaves as a metal. If the anode can be attacked by the anion it is dissolved into the solution; this is the case in E, in which the copper sulphate $CuSO_4$ is decomposed into the kation Cu and the anion SO_4. The latter attacks the copper of the anode, which it dissolves, leaving impurities in the copper behind as scale. In the vessel F the same decomposition takes place and the copper appears at the kathode. The SO_4 combines with the hydrogen of the water H_2O to form H_2SO_4 (sulphuric acid) and the oxygen is set free at the anode. In the third beaker G the products are, oxygen at the anode, and silver at the kathode.

116. Faraday's laws of Electrolysis. These laws are two in number, and connect together the amount of the ions deposited by a current in different electrolytes and the quantity of electricity which passes.

LAW I. *The mass of an electrolyte set free by the passage of a current of electricity is directly proportional to the quantity of electricity which has passed through the electrolyte.*

Thus if m grammes of a substance be deposited by the passage of q units of electricity, then m is proportional to q; in other words, the ratio m/q is constant for that substance.

LAW II. *If the same quantity of electricity passes through different electrolytes, the masses of the different ions deposited will be proportional to the chemical equivalents of the ions.*

Thus if the same current pass through acidulated water, copper sulphate, and silver nitrate in succession, then for each gramme of hydrogen collected there will be 8 grammes of oxygen, 31·6 of copper, and 108 of silver.

Moreover it is a consequence of the first law that the

mass deposited depends on the quantity of electricity which has passed, and not on the strength of the current. A weak current flowing for a long time produces the same deposit as a stronger current flowing for a shorter time, provided the quantity of electricity transferred is the same in the two cases.

It should be noted however that in consequence of various secondary actions the nature of the deposit depends in some cases on the rate at which it takes place; if this be too great the deposit does not adhere to the kathode.

117. Electro-chemical Equivalent. We can state the laws more concisely by the introduction of the term electro-chemical equivalent of a substance.

DEFINITION OF ELECTRO-CHEMICAL EQUIVALENT. *The* **Electro-chemical Equivalent** *of a substance is the number of grammes of that substance deposited during the passage of a unit quantity of electricity.*

Let the electro-chemical equivalent of a substance be γ, and let m grammes of the substance be deposited by the transference of q units of electricity.

Then since γ grammes are deposited during the transference of each unit of electricity γq grammes are deposited by the transference of q units.

Hence $\qquad\qquad m = \gamma q.$

Moreover if the q units are transferred by the passage of a uniform current of strength c flowing for t seconds, we have $q = ct.$

Hence. $\qquad\qquad m = \gamma ct.$

From this we have

$$\gamma = \frac{m}{ct}.$$

If then we can measure the current, the time during which it has been flowing, and the mass of the substance deposited, we can find γ, its electro-chemical equivalent.

If this be done for a number of substances, it will be found that the electro-chemical equivalents of the various substances are proportional to their chemical equivalents, and this is what is stated in Faraday's second law.

It must be remembered that in different cells of the series each atom of a given element may take the place of one or more atoms of hydrogen. Thus the atomic weight of copper is 63·2 but in copper sulphate, $CuSO_4$, each atom of copper is equivalent to two atoms of hydrogen, thus the chemical equivalent of copper in copper sulphate is 63·2/2 or 31·6. Calling the number of atoms of hydrogen which are replaced by a given element in a given combination its "valency" in that combination, then the chemical equivalent is the atomic weight divided by the valency. Hence if we know the electro-chemical equivalent of hydrogen, we can find that of any other element by multiplying the chemical equivalent of that element by the electro-chemical equivalent of hydrogen.

The electro-chemical equivalent of hydrogen will depend on the unit we choose to measure our "unit quantity of electricity."

If we suppose that our unit current is one ampere, the unit quantity of electricity is the quantity conveyed by one ampere flowing for one second, and it has been shewn by direct experiment that this quantity deposits ·00001038 grammes of hydrogen.

From this we can get the electro-chemical equivalent of any other substance. Some of these are given in the following Table.

Element	Chemical Equivalent	Electro-Chemical Equivalent
Electro-positive		
Hydrogen	1	·00001038
Sodium	23	·0002388
Silver	108	·001118
Copper Cupric	31·6	·0003281
Copper Cuprous	63·2	·0006562
Zinc	32·5	·0003370
Lead	103·2	·001071
Electro-negative		
Oxygen	8	·00008286
Nitrogen	4·66	·00004849
Iodine	127	·001314

The numbers in the third column are in all cases obtained by multiplying those in the second by the electro-chemical equivalent of hydrogen.

The elements are classed as (1) electro-positive, those which are carried forward by the current and appear at the kathode, and (2) electro-negative, those which are as it were left behind, being attracted to the anode at which the current enters the liquid.

The reasons for the names will appear more clearly when we deal with the theory of electrolysis.

Since one ampere is one-tenth of the c.g.s. unit of current (see Section 148), in order to find the quantity deposited by the passage of 1 c.g.s. unit of electricity we must multiply each of the numbers in the Table by 10.

118. Local Action. When a piece of pure zinc is immersed in dilute acid there is no action between the two ; with ordinary commercial zinc hydrogen is given off and the zinc is dissolved. This is due to the fact that commercial zinc is impure, containing among other things iron. Now the iron, the zinc, and the acid combined form a small local battery, a current passes from the iron to the zinc and back to the iron through the acid. This current causes electrolysis of the acid and the liberation of hydrogen. The SO_4, which is also set free, attacks the zinc, and forms sulphate of zinc, thus the zinc is dissolved and hydrogen produced. When commercial zinc is used in a battery this action is set up, and unless steps are taken to prevent it the zinc is being continually dissolved, even when the battery is on open circuit. This action known as "local action" is remedied by amalgamating the zinc with mercury ; the surface of the zinc is first cleaned with acid and then a few drops of mercury are run on to it and rubbed in with a rag or stick. The mercury and zinc form a pasty mass in which the particles of iron or other impurity float. When local action is set up the iron is carried off by the hydrogen bubbles as they rise to the surface of the liquid. Thus the cause of the action is removed and a clean bright surface of zinc amalgam left for the acid to act upon.

119. Voltaic Cells. We have seen that when a piece of zinc and a piece of copper are placed in dilute acid and connected by copper wires to the opposite quadrants of an electrometer a difference of potential, which we have called the electromotive force of the battery, is indicated by the electrometer, while if the plates be connected by a wire the potential difference falls and a positive current passes through the wire from the copper to the zinc and back through the acid from the zinc to the copper. Now the passage of a current through the acid electrolyses it. Hydrogen is carried by the current to the copper plate and deposited on it, while the sulphion SO_4 formed on the zinc plate, attacks the zinc forming zinc sulphate $ZnSO_4$. The process may be represented by the equation

$$Zn + H_2SO_4 = ZnSO_4 + \overset{\bullet}{H}_2.$$

Zinc and sulphuric acid combine to produce zinc sulphate and hydrogen.

Moreover the quantities of the substances involved in these changes are chemically equivalent. For each gramme of hydrogen set free, we have liberated 32/2 or 16 grammes of sulphur, and $16 \times 4/2$ or 32 grammes of oxygen, forming sulphion, and these combine with 65/2 or 32·5 grammes of zinc to form $16 + 32 + 32·5$ or 80·5 grammes of zinc sulphate. This process continues but the current falls off, for instead of having a copper plate in the acid, we have a copper plate coated with hydrogen; the effect of this hydrogen is to set up a current in the opposite direction to that of the battery, and hence the original current is reduced. This effect is said to be due to polarization, and the battery is said to be polarized.

We have already seen that a current in a wire exerts magnetic force in its neighbourhood, an instrument arranged to make use of this magnetic effect to measure the current is called a galvanometer. See Section 149. By the aid of a galvanometer we can detect this reverse current due to polarization thus:

EXPERIMENT 33.   *To shew the reverse current due to polarization.*

A battery of two or more Daniell cells, a water voltameter, consisting of two platinum plates immersed in a beaker of slightly acidulated water, and a delicate galvanometer are connected as shewn diagrammatically in Fig. 124.

The handle S of a switch is connected to the galvanometer G, the other terminal of the galvanometer is connected to one plate of the voltameter V, and the second plate of the voltameter is connected to the battery B. This point of junction is connected to one terminal R of the switch, the other terminal T is joined to the second pole of the battery.

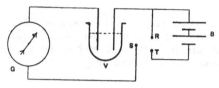

Fig. 124.

Thus when the handle of the switch connects S and T a circuit is complete through the battery, voltameter and galvanometer. When the handle is moved across to S the battery is cut out of the circuit, which is complete through the voltameter and galvanometer.

Place the switch so that S and T are connected. A current flows round the circuit, electrolysis goes on in the voltameter, gases being deposited on the plates, and the needle of the galvanometer is deflected. Note the direction of the deflexion. Then reverse the switch so that S and R are connected, the battery is now out of circuit; the galvanometer however still shews a current, but in the opposite direction to that which previously deflected it. The plates are polarized and an E.M.F. of polarization is acting. The oxygen and hydrogen of the voltameter in part recombine, and as this combination becomes more complete the current dies away to zero.

This same process occurs in the simple voltaic cell when it is producing a current; in consequence of the opposing E.M.F. of polarization the electromotive force of the cell falls rapidly with use and the current diminishes. For this reason various modifications of the cell have been devised which aim at reducing the polarization.

In most cells the negative plate is made of zinc, the positive plate varies; in all cases however hydrogen tends to pass from the liquid near the zinc to the positive plate, and the object aimed at is to neutralize the consequences of its deposition. This result is achieved by putting an oxidizing

agent either into the liquid or into the material of the
positive plate.   The electromotive forces of the different
cells—measured it will be remembered by the differences of
potential between their plates when on open circuit—are
different.

In some cells a single fluid is used, in others the cell is
divided into two compartments separated by a plate of porous
earthenware, the zinc plate is in one compartment, the positive
plate in the other, and the two compartments contain different
fluids.

120. Single fluid cells.   In these the oxidizing agent
may either be in the fluid or in the positive plate.   As
oxidizing agents various substances might be used, as nitric
acid, bichromate of potash, black oxide of manganese, or
peroxide of lead.

Most of these however would attack the copper, and so
the copper has to be replaced by some substance which will
resist the chemical action of the oxidizer.   Carbon and
platinum are such substances and are in consequence used
in various cells; thus we might have a cell containing zinc
and carbon in nitric acid for example.   But nitric acid would
dissolve the zinc on open circuit.   Hence it cannot be used in
the same vessel as the zinc plate; we can only employ it in a
two-fluid cell.

Bichromate of potash however dissolved in sulphuric acid
is used in the bichromate cell, though in this case also it is
necessary to provide means for withdrawing the zinc plate
when the cell is out of action.   In the Leclanché and dry
cells, binoxide of manganese is mixed with the carbon of the
positive plate, the exciting agent being chloride of ammonia.

121. The Bichromate Cell.   This is usually made
up in the bottle form shewn in Fig. 125.   The positive pole
is composed of two plates of hard carbon connected together
and to one binding screw.   The negative pole is a plate of
zinc which is connected to the other binding screw, and can
be withdrawn from the liquid as shewn in the figure.   The
liquid is a solution of bichromate of potash in sulphuric acid.
The electromotive force of the cell on open circuit is about

2·1 volts. When the poles are connected by a wire a current flows in the wire from the carbon to the zinc. The sulphuric acid is electrolysed and zinc sulphate formed, but the hydrogen combines with some of the oxygen of the bichromate and polarization is prevented. In consequence of the proximity of the plates in the acid solution the internal resistance (§ 135) of this cell is low, and it is possible to use it to give a large current through a suitable external circuit.

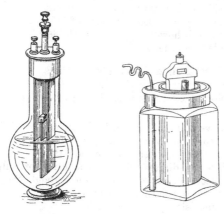

Fig. 125.   Fig. 126.

122. Leclanché Cell. In this cell, which in its usual form is shewn in Fig. 126, the liquid is salammoniac, the negative plate is a rod of amalgamated zinc, and the positive plate a mixture of carbon and black oxide of manganese. This mixture is pounded up and then compressed into a hard cake; in the older form of cell it is put inside a porous pot. When in use the cell polarizes; a double chloride of zinc and ammonia is formed, and hydrogen and ammonia are set free which collect on the carbon. If the cell however be only used for a short period, and then left on open circuit, the manganese binoxide gradually gives off oxygen, which combines with the hydrogen and depolarizes the positive plate. The cell is thus very convenient for use on telegraph circuits, electric bells, telephones, and the like where it is only wanted for a short interval at a time and then has a period of rest. It is easily set up, is clean, and requires very little attention. Its electromotive force is about 1·35 volts.

123. The Dry Cell. This, which is shewn in Fig. 127, is a modification of the Leclanché arranged to secure portability. The active materials are the same, but the cell is filled with a paste of sulphate of lime, or some such material, soaked in chloride of ammonia. This pasty mass carries the current in the same way as the solution of the Leclanché, the cell however is more portable, and less liable to accident.

124. Two-Fluid Batteries. These are designed to counteract the direct action of the depolarizer on the zinc of the negative pole.

The positive plate is immersed in a strong depolarizing solution in a porous pot. This is put inside another vessel, which contains the zinc plate in dilute acid or a solution of sulphate of zinc. The electrical action can go on through the pores of the pot, but the depolarizer can only reach the zinc slowly by diffusion through the pores. Sometimes the zinc and acid are in the porous pot, the positive plate being in the outer vessel, but this of course makes no difference to the action.

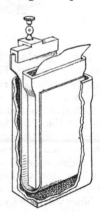

Fig. 127.

125. Grove's Cell. This is shewn in Fig. 128. The positive plate consists of a piece of platinum foil; this is immersed in strong nitric acid. The porous pot is usually flat in shape, and the negative plate is a zinc sheet bent so as to encompass the pot closely; both are placed in a porous pot which contains sulphuric acid; thus the plates are near together and the internal re-

Fig. 128.

sistance, § 135, is low. The sulphuric acid is decomposed, forming zinc sulphate and hydrogen; the latter passing through the pot is oxidized by the nitric acid which is itself reduced. The electromotive force is about 1·95 volts.

126. Bunsen's Cell. This cell is similar to the Grove cell, only the platinum, which is costly, is replaced by gas carbon. Its action and electromotive force are both the same as that of Grove's cell.

127. Daniell's Cell. In this cell, Fig. 129, the positive plate is copper immersed in sulphate of copper, the negative plate, zinc in sulphuric acid. The two liquids are kept apart by a porous partition; in the form shewn in the figure the copper plate is placed in the outer vessel of the cell, the zinc plate and sulphuric acid are contained in the porous pot which is placed inside the copper vessel.

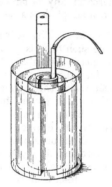

The sulphuric acid is decomposed, forming zinc sulphate and hydrogen, the hydrogen traverses the porous pot and replaces the copper in the copper sulphate, forming sulphuric acid and copper, and the copper is deposited on the copper plate. Thus the nature of that plate is not changed by the

Fig. 129.

passage of the current and there is no polarization. In order to maintain the copper sulphate solution of constant strength crystals of sulphate of copper are placed in the solution, usually in a small tray at the top of the cell, these are gradually dissolved, thus replacing the copper deposited on the copper plate.

The chemical action may be represented by the two following equations :

$$Zn + H_2SO_4 = ZnSO_4 + H_2.$$

Zinc and Sulphuric Acid give Zinc Sulphate and Hydrogen.

$$H_2 + CuSO_4 = H_2SO_4 + Cu.$$

Hydrogen and Copper Sulphate give Sulphuric Acid and Copper.

Sometimes the cell is made up with a concentrated solution of zinc sulphate in water, instead of the sulphuric acid. The electromotive force depends on the solution, and varies from about 1·18 volts when sulphuric acid diluted with twelve parts of water is used, to 1·07 volts when zinc sulphate is used.

For solutions of a given strength, however, the electromotive force of the cell is very constant.

128. Standards of Electromotive Force. A cell of constant electromotive force may conveniently be employed as a standard in terms of which to measure the electromotive force of any other cells. For this purpose the Clark cell has been adopted as a legal standard in many countries.

129. The Clark Cell. This may take various forms. One such form is shewn in Fig. 130. The cell is contained in a glass test-tube. The positive pole is pure mercury, and communication is made with this by means of a platinum wire which passes through a glass tube into the mercury. Above the mercury is a paste of mercurous sulphate dissolved in pure zinc sulphate; this is covered again by a saturated solution of zinc sulphate containing crystals of zinc sulphate so as to remain saturated at any temperature at which the cell may be used. An amalgamated rod of pure zinc dips into the zinc sulphate and forms the negative pole. The cell is closed with a cork and sealed with marine glue. It is not to be used as a source of a current, for it will polarize, but merely as a standard of electromotive force on open circuit, or in such

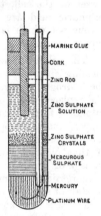

Fig. 130.

circumstances that the current which can be formed must be infinitesimal. The electromotive force like that of other cells depends on the temperature. At 15° C. its value is 1·434 volts[1].

---

[1] More recent experiments appear to shew that this number should be reduced probably to 1·4328, but the question is now under investigation.

Another form of Clark cell which has many advantages is shewn in Fig. 131. The positive pole is mercury contained in one of the two test-tubes, the negative pole an amalgam of zinc and mercury. Communication is made with the poles by means of platinum wires sealed through glass tubes.

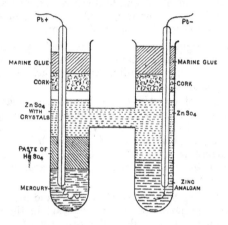

Fig. 131.

Above the mercury is the mercurous sulphate paste, above the zinc the saturated zinc sulphate solution containing crystals of the salt. Communication between the two goes on through the horizontal tube which is filled with zinc sulphate. This pattern, known as the *H* form of cell, was devised by Lord Rayleigh. The materials are more completely separated than in the other pattern and the E.M.F. is more constant.

130. Weston Cell. For some purposes Weston's modification of the Clark cell is very useful. In place of the zinc Weston uses cadmium immersed in cadmium sulphate; the mercurous sulphate paste is also made up with cadmium sulphate. The great advantage of the cell is that its E.M.F. varies very little with temperature; its value is 1·018 volts at temperatures near 15° C.

131. Secondary Batteries. We have already seen

that when two platinum plates are immersed in dilute acid they become polarized on the passage of a current, and will, if connected directly, give a current for a brief time during which the polarization is reduced. Planté shewed that if the platinum plates be replaced by lead plates treated in a certain manner by the passage of a current backwards and forwards it becomes possible to store up in the cell a quantity of electricity and to utilize the cell as a source of current.

Planté's original process was modified by Faure. Two plates of lead A and B, coated with minium or red lead, are placed in a cell containing dilute sulphuric acid and a current is passed from A to B.

The red lead on A becomes peroxidized, that on B is reduced to a lower oxide, and then finally to the condition of spongy metallic lead. If these two plates are now connected together, the original source of current being removed, a reverse current is produced passing from B to A in the cell and from A to B in the wire, and this goes on until the original condition is reached.

By continuing the process of charging and discharging for some time the amount of lead taking part in the changes is gradually enlarged, and thus the capacity of the cell, as measured by the quantity of electricity it can hold before the hydrogen begins to come off in bubbles from the plate B, is considerably increased.

In the more modern form of accumulator or storage cell the plates take the form of a grid of metallic lead into the interstices of which a paste of red lead and sulphuric acid is pressed. The cell is formed by the passage of a current which peroxidizes the paste on one plate and reduces it on the other.

When the cell is in use the specific gravity of the acid solution should be about 1·18. The electromotive force of the cell is about 2 volts, and continues at this value until the cell is almost completely discharged; as this stage approaches a very rapid fall in the electromotive force is observed; the discharge should be stopped when the E.M.F. reaches 1·85 volts and the cell recharged.

In practice the capacity of an accumulator is measured in ampere-hours, an ampere-hour being the quantity of electricity conveyed by a current of 1 ampere flowing for one hour. It is thus 3600 coulombs.

Accumulators usually contain a number of plates ranged side by side, the odd plates are connected together to form one pole of the cell, while the even plates connected together form the other. The capacity of the cell depends on the size and number of its plates, its electromotive force is independent of this, and is determined only by their nature and the state of their surfaces. Fig. 132 shews a form of secondary cell in general use.

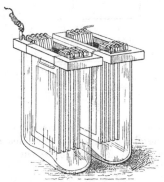

Fig. 132.

132. Arrangement of Batteries in Series. Consider two batteries A, B (Fig. 133). Let C_1, Z_1 be the copper and zinc plates of the one, C_2, Z_2 those of the second; E_1 and E_2 being the electromotive forces. Connect together Z_1 and C_2. Connect C_1 and the junction of Z_1 and C_2 by two wires to an electrometer; a potential difference E_1 will be observed, the potential of the quadrant in connexion with C_1 being higher than that of the quadrant connected to the junction by an amount E_1. Now connect the junction and Z_2 to the electrometer; the potential of the quadrant connected to the junction will be found to exceed by E_2 that of the quadrant connected to Z_2. Thus the potential difference between C_1 and Z_2 is $E_1 + E_2$, and this may be verified by connecting C_1 and Z_2 to the electrometer.

Two batteries arranged so that the negative pole of the one is connected to the positive pole of the other are said to be connected in series, and the electromotive force of such an

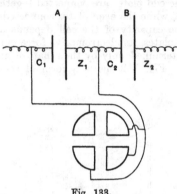

Fig. 133.

arrangement is the sum of the electromotive forces of the two batteries. In general if a number of batteries are so connected the same law holds; if E_1, E_2 etc. be the individual electromotive forces, then

$$E = E_1 + E_2 + E_3 + \ldots\ldots$$

133. Arrangement of Batteries in Multiple Arc or Parallel. Consider now two cells having the same E.M.F., E. The potential difference both between C_1 and Z_1, and between C_2 and Z_2, is E. If then C_1 and C_2 (Fig. 134) be connected by one wire to one pair of quadrants of an electrometer, and Z_1 and Z_2 be connected by a second wire to the other pair of quadrants, the electrometer will still indicate a potential difference E. The cells are said to be connected in parallel or in multiple

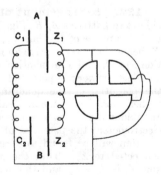

Fig. 134.

arc, and when the two have the same E.M.F. this is also the E.M.F. of the combination.

If the two cells be not equal in E.M.F. the stronger cell will send a current round the circuit and the problem of finding the potential difference between the plates is more complex.

CHAPTER XIV.

RELATION BETWEEN ELECTROMOTIVE FORCE AND CURRENT.

134. Electromotive Force and Current in a Simple Circuit. When a difference of potential is established between two points on a conductor a current flows in the conductor; it remains for us to consider the relation between the strength of this current and the difference of potential or electromotive force.

Now experiment proves that for a conductor composed of a single material in a given physical condition the ratio of the difference of potential between two points to the current flowing between these points is a constant. This constant is known as the **Resistance** of the conductor. Let us denote this quantity by R. Let V_1, V_2 be the potentials between the points and let C be the current flowing between them. Then we have the result that

$$\frac{V_1 - V_2}{C} = R,$$

where R, the resistance, is a constant quantity depending on the shape, material, and other physical properties of the conductor, but not on the current or on the difference of potential.

We can of course put the equation into the form

$$C = \frac{V_1 - V_2}{R}.$$

As will be seen in the sequel the resistance of a conductor depends on its temperature, and since the passage of the current heats the conductor its resistance is to this extent dependent on the current.

The experimental verification of this result will come at a later stage (see § 163). For the present we will consider some further consequences of it.

It follows from the above that any conductor has a definite resistance—we shall see later how this may be measured—and that if we know the resistance we can calculate the current produced by a given difference of potential applied to the ends of the conductor, or conversely the difference of potential required to produce a given current.

135. Electromotive Force and Current in a Circuit containing a Battery. The above statement assumes that we have no source of electromotive force in the circuit between the points at which the potentials V_1, V_2 are measured; the conditions correspond to those which hold in the case of water flowing down a sloping tube. The flow will depend on the nature of the tube and on the difference of pressure between the ends.

The circumstances are entirely altered if in the tube we place a turbine or wheel tending to propel the water in the tube either up or down. If there be a battery or some other source of electromotive force between the points at which the potentials are measured then the E.M.F. of this battery must be added to, or subtracted from, the potential difference between the points in order to get the resultant electromotive force to which the current is due. Thus consider a circuit consisting of a battery of E.M.F. E, and resistance R_2 the poles of which are connected by a wire of resistance R_1. The quantity E it will be remembered is measured by the difference of potential between the poles of the battery when on open circuit.

Let V_1 be the potential of the positive pole, V_2 that of the negative pole. Then as we pass along the wire from the copper to the zinc, in the direction that is in which the current is flowing, the electromotive force is $V_1 - V_2$, and the resistance R_1, hence if C be the current

$$C = \frac{V_1 - V_2}{R_1},$$

or $V_1 - V_2 = CR_1$.

But as we pass from the zinc to the copper through the battery, still going with the current, the electromotive force is $V_2 + E - V_1$, the resistance is R_2, and the current C.

Hence if we suppose the same law to hold

$$C = \frac{V_2 + E - V_1}{R_2},$$

or $$E - (V_1 - V_2) = CR_2;$$

adding these two results we find

$$E = C(R_1 + R_2) = CR$$

if $$R = R_1 + R_2,$$

and this may be written in the alternative forms

$$\frac{E}{C} = R,$$

or $$\frac{E}{R} = C.$$

In the above case R is defined as the resistance of the circuit.

136. Ohm's Law. When the current is due to the action of a battery we can to a certain extent localize the seat of the E.M.F.—it is at any rate somewhere in the battery though it may be difficult to say at exactly what point it acts; there are however many cases in which the electromotive force acts continuously at all points of the circuit, still experiment shews that in all cases the law connecting the three quantities E.M.F., current, and resistance is the same. This law, which is called from its discoverer Ohm's law, may be stated thus:

OHM'S LAW. *In any circuit the ratio of the electromotive force producing a current to the current produced is a constant depending only on the form, materials, and physical conditions of the circuit. This constant is called the* Resistance *of the circuit.*

We may express Ohm's Law in symbols thus:—If E be the electromotive force, C the current, and R the resistance, then

$$\frac{E}{C} = R,$$

and R is constant for the circuit.

137. Unit of Resistance. Electrical resistance, like other quantities, is measured in terms of a proper unit, and this unit is called an "**Ohm**."

The definitions of the Ohm, the Ampere, and the Volt, are based ultimately on certain theoretical considerations, and are connected together in such a way that an ampere is the current produced by one volt acting through a resistance of one ohm. From this it follows that if, in the equation representing Ohm's law, we measure E and R in volts and ohms, then C is measured in amperes.

Thus for example if the resistance of a circuit is 50 ohms and an E.M.F. of 5 volts acts round it, the current is 5/50 or 1/10 of an ampere ; or again, if the current is 5 amperes and the resistance 50 ohms, the E.M.F. is 5 multiplied by 50 or 250 volts.

Since the resistance of a conductor is a physical property of the conductor depending on its material, shape and conditions, any given conductor, such as a piece of wire, has a definite resistance in ohms.

A column of mercury 106·3 centimetres in length, and 1 square millimetre in cross section, has been found to have, when at the temperature of melting ice, a resistance very approximately equal to that of the ohm as theoretically defined, and it has been agreed to take the resistance of such a column as the practical unit of resistance and call it one ohm.

In order to get over certain difficulties of measurement, the column is defined by its length, and the mass of mercury it contains at zero Centigrade, instead of by its length and cross section. Accordingly the following practical definition has been generally agreed upon.

DEFINITION. *A column of mercury of uniform cross section, 106·3 centimetres in length, which contains at the temperature of melting ice a mass of* 14·4521 *grammes of mercury has a resistance of an* **Ohm**.

An electromotive force of one volt, applied to the ends of such a column, will produce a current of one ampere.

138. Conductance. If R be the resistance of a conductor, the quantity $1/R$ is known as its **Conductance**.

It is the ratio of the current in the conductor to the electromotive force producing it. The conductance of a conductor whose resistance is 1 ohm is unity, that of a conductor whose resistance is 50 ohms is 1/50, and so on.

139. Conductors in series.

PROPOSITION 13. *To find the resistance of a number of conductors in series.*

Let A_1A_2, A_2A_3, A_3A_4, etc., Fig. 135, be a series of conductors connected together at A_2, A_3, etc. Let V_1, V_2, V_3, ... be the potentials at A_1, A_2 ..., R_1, R_2, ... the resistances of the conductors, and let a current C traverse this series.

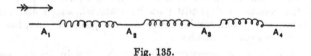

Fig. 135.

Then we have
$$V_1 - V_2 = CR_1$$
$$V_2 - V_3 = CR_2$$
$$V_3 - V_4 = CR_3$$
$$\dots\dots\dots\dots$$
$$V_n - V_{n+1} = CR_n.$$

Hence adding these
$$V_1 - V_{n+1} = C\ (R_1 + \dots R_n).$$

But if R denotes the resistance between the extreme points A_1 and A_{n+1} between which the current C is flowing, then
$$V_1 - V_{n+1} = CR.$$

Hence
$$R = R_1 + \dots R_n.$$

In other words the resistance of a number of conductors connected so that the same current traverses each in turn, is the sum of the resistances of the various conductors.

It follows clearly from this that the resistance of a piece of wire of uniform material and thickness, is proportional to the length of the wire. For let the wire be l cm. in length, we may consider it as l equal conductors in series, each of them being 1 cm. in length. Let σ be the resistance of 1 cm. of the wire, then the resistance of the l conductors in series, each of a resistance σ, is $l\sigma$. Thus the whole resistance is $l\sigma$; it is therefore proportional to the length of the wire.

140. Conductors in parallel. When two or more conductors join the same two points A, A', so that a current from A to A' can flow by two or more paths, the conductors are said to be in parallel, or sometimes in multiple arc.

PROPOSITION 14. *To find the resistance of a number of conductors in parallel.*

Let AA_1A', AA_2A', etc., Fig. 136, be a number of conductors joining two points A and A'. Let a current C be led

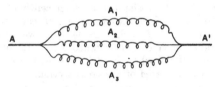

Fig. 136.

into the system at A and withdrawn at A', and let C_1, C_2, C_3 be the currents in the various conductors, R_1, R_2, etc. the resistances of the conductors. Let V, V' be the potentials at A and A'. Now the sum of the currents in the various paths between A and A' is equal to the current led in at A.

Hence $\qquad C = C_1 + C_2 + \ldots + C_n$.

Also $\qquad\qquad C_1 = \dfrac{V - V'}{R_1}$,

$$C_2 = \frac{V - V'}{R_2},$$

etc.

Hence

$$C_1 + C_2 + \ldots = (V - V')\left(\frac{1}{R_1} + \frac{1}{R_2} + \ldots + \frac{1}{R_n}\right).$$

And if R is the equivalent resistance, the resistance that is of a conductor which under the potential difference $V - V'$ will permit of the passage of the current C,

then
$$C = \frac{(V - V')}{R}.$$

Hence
$$\frac{1}{R} = \frac{1}{R_1} + \frac{1}{R_2} + \dots + \frac{1}{R_n},$$

or in words:

The conductance of a number of conductors in parallel is the sum of the conductances of the several conductors.

It follows from this that the conductance of a wire is proportional to the area of its cross section. For consider a wire of any length l cm. and a sq. cm. in area of cross section. We may split it up into a wires each 1 sq. cm. in area placed side by side. Let the resistance of each centimetre of each of these wires be ρ, then the resistance of each wire is ρl, and the conductance of each wire is $1/\rho l$; we have now a wires in parallel, each of conductance $1/\rho l$. It follows that the conductance of the whole is $a/\rho l$; that is, the conductance is proportional to the area of the cross section.

If R be the resistance of the wire, then we have seen that

$$\frac{1}{R} = \frac{a}{\rho l}.$$

Hence
$$R = \frac{\rho l}{a}.$$

141. Specific Resistance[1]. Thus the resistance of a uniform wire is proportional to its length and inversely proportional to the area of its cross section. The quantity ρ is the resistance of a piece of the wire 1 cm. in length, having an area of 1 sq. cm. in cross section : the form of this cross section is not material; if we imagine it to be square we see that ρ is the resistance between two opposite faces of a cube each edge of which is 1 centimetre. This quantity is called the **Specific Resistance** of the material of the wire.

[1] It should be noticed that the term specific is used in its proper sense and does not imply a ratio as in " specific " heat, "specific" gravity.

DEFINITION. *The* **Specific Resistance** *of a material is the resistance between two opposite faces of a cube of the material each edge of which is* 1 *centimetre in length.*

Thus we have the result that if ρ be the specific resistance of a wire of length l cm. and cross section a sq. cm., and R the resistance of the whole wire,

then $$R = \frac{\rho l}{a} \text{ ohms.}$$

Conversely if we are given the resistance, length and cross section of a wire we can find the specific resistance of its material from the formula

$$\rho = \frac{Ra}{l} .$$

The following Table gives the specific resistance of a number of materials in microhms[1] per cube centimetre and the resistance in ohms of a length of 1 metre, 1 square millimetre in cross section at 0° C.

TABLE.

Material	Specific Resistance in Microhms	Resistance of 1 metre 1 sq. mm. in area of cross section in Ohm
Silver annealed	1·468	·01468
,, hard drawn	1·615	·01615
Gold annealed	2·036	·02036
Zinc	5·751	·05751
Copper annealed	1·562	·01562
,, hard drawn	1·603	·01603
Iron	9·065	·09065
Platinum	10·917	·10917
Mercury	94·073	·94073
Platinum Silver	24·120	·24120
German Silver	20·243	·20243
Manganin	46·700	·46700
Resista	76·490	·76490

142. Distribution of current between a number of conductors in parallel. The question may often arise as to how a current will distribute itself in a series of conductors in parallel. The principles of the foregoing proposition enable us to answer this.

---

[1] A microhm is a millionth of an ohm or 10^{-6} ohms.

PROPOSITION 15. *To find the distribution of current in a number of conductors in parallel.*

As in Fig. 136 above, let AA_1A', AA_2A', etc. be the conductors. Let C be the current entering at A and leaving at A'; while C_1, C_2, $C_3 \ldots$ are the currents, R_1, R_2, $R_3 \ldots$ the resistances of the conductors. Then since the potential difference between A and A' is the same for each of the possible paths, and since the potential difference is measured by the product of the current and the resistance we must have

$$C_1R_1 = C_2R_2 = C_3R_3 = \ldots$$

and

$$C = C_1 + C_2 + C_3 + \ldots.$$

Hence
$$C = C_1R_1 \left\{ \frac{1}{R_1} + \frac{1}{R_2} + \ldots \right\} = \frac{C_1R_1}{R}.$$

Thus
$$C_1 = \frac{CR}{R_1},$$
$$C_2 = \frac{CR}{R_2},$$
$$\text{etc.}$$

where $1/R$ is the conductance of the system, being given by the equation

$$\frac{1}{R} = \frac{1}{R_1} + \frac{1}{R_2} + \ldots \frac{1}{R_n}.$$

If there be two circuits only of resistances R_1 and R_2

$$\frac{1}{R} = \frac{1}{R_1} + \frac{1}{R_2},$$
$$R = \frac{R_1R_2}{R_1 + R_2}.$$

Hence
$$C_1 = \frac{CR_2}{R_1 + R_2},$$
$$C_2 = \frac{CR_1}{R_1 + R_2}.$$

This result is important in the theory of galvanometers, § 159.

143. Graphic representation of Ohm's Law.

Let us suppose that on a diagram we represent potential

differences by vertical straight lines, and resistances by horizontal straight lines. Thus in Fig. 137, let PM represent the electromotive force E round a given circuit, and let MN represent R the resistance of the circuit.

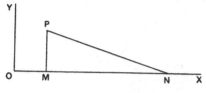

Fig. 137.

Then the current C which is given by the ratio E/R is represented by PM/MN or by $\tan \widehat{PNM}$.

144. Chemical Theory of a Voltaic Cell. If the circuit consist of a series of different materials, and we know how the potential changes as we pass from one material to the next we can draw a similar diagram. Thus in the case of a voltaic cell, there is in the views of many electricians reason to suppose that when the zinc is dipped into the dilute acid, its potential is less than that of the acid by 1·8 volts, while the potential of the copper falls below that of the acid by about ·8 of a volt. If a piece of copper wire be connected to the zinc it will be at the same potential as the zinc. Thus when the circuit is open the distribution of potential is as shewn in Fig. 138.

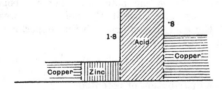

Fig. 138.

If the circuit be completed by joining the free end of the copper wire to the copper plate we proceed thus :

Let OA, Fig. 139, represent the resistance of the zinc, AB of the acid, BC of the copper, CD of the wire. Draw ON perpendicular to $OABCD$ to represent E the E.M.F. of the cell and join ND. Then the current is represented by the tangent of the angle NDO.

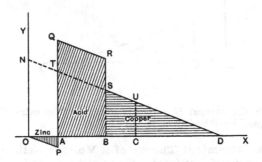

Fig. 139.

Let PAQ and BSR be drawn parallel to ON. Draw OP parallel to ND to meet PAQ in P, and take PQ to represent the difference of potential between the zinc and the acid. From Q draw QR parallel to ND to meet BSR in R.

Let PQ meet ND in T.

Then $RS = QT$.

Also $PT = ON = $ E.M.F. of cell.

But PQ is the potential difference between the zinc and the acid, hence QT is the potential difference between the acid and the copper. Thus if the point R represents the potential of the acid in contact with the copper, S represents the potential of the copper.

Hence if the point O represents the potential of the point of contact of the copper and zinc, the curve $OPQRSD$ represents the distribution of potential.

The current which is uniform throughout is represented by the tangent of the inclination of the parts OP, QR, SD to the resistance line; the potential falls uniformly along OP, through the zinc, then there is a sudden rise at the zinc surface to PQ, a uniform fall from AQ to BR through the acid, a sudden drop RS at the copper surface, and a uniform fall along SD from SB to the original value, through the copper and copper wire.

Moreover let CU parallel to ON meet SD in U; if the ends of the external circuit represented by C and D be connected to an electrometer, the potential difference registered will be represented by CU. When the battery was on open circuit the potential difference was ON the E.M.F. of the battery, and we see hence how it is that the difference of potential between the plates of a battery falls, as stated in Section 107, when the circuit is closed through an external resistance.

If we call E the E.M.F. of the battery, E_1 the potential difference between the ends of the external resistance, R the whole resistance, and R_1 the external resistance, then in the figure

$$ON = E, \qquad CU = E_1,$$
$$OD = R, \qquad CD = R_1,$$

and OC or $R - R_1$ is the battery resistance.

Then we have clearly

$$\text{Current} = \tan NDO$$
$$= \frac{E}{R} = \frac{E_1}{R_1} = \frac{E - E_1}{R - R_1}.$$

145. Contact theory of a Voltaic cell. According to another theory of the cell, the copper, zinc, and acid, when on open circuit are all at very nearly the same potential, but if a copper wire be connected to the zinc, a difference of potential is established between the zinc and the copper, the potential of the zinc exceeding that of the copper by about 1 volt, and this measures the electromotive force of the cell.

In this case the open circuit distribution is as in Fig. 140. When the free end of the copper wire is connected to the copper plate its potential is raised to that of the plate and a current runs through the wire from the copper to the zinc; the distribution of potential is as in Fig. 141, where O represents the potential of the copper wire at its junction with the zinc,

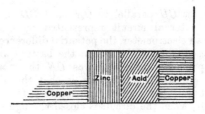

Fig. 140.

N is the potential of the zinc where it is joined by the wire, and the fall of potential through the cell and wire is given by the straight line $NQRUD$.

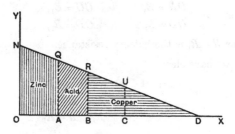

Fig. 141.

In this case also we clearly have, as before,

$$\text{Current} = \frac{E}{R} = \frac{E_1}{R_1} = \frac{E - E_1}{R - R_1}.$$

The theory of the cell and the experiments on which it is based are given in Sections 191--196.

CHAPTER XV.

MEASUREMENT OF CURRENT.

146. Measurement of a Current. Galvano-meters. A current may be measured either by its magnetic, its chemical, or its thermal effects. Various instruments have been devised for utilizing the magnetic force due to a current to measure it. These are called galvanometers.

We have seen that there is magnetic force exerted in the neighbourhood of a wire carrying a current; if we bring such a wire near a compass-needle it is in general deflected; the deflexion is increased by bringing the needle near to the wire; it is also increased by strengthening the current. Bend the wire into the form of a circle some 8 or 10 cms. in radius, and hold it in a vertical plane so that the point of support of the compass-needle is at its centre and the plane of the coil is north and south, the compass-needle is deflected by the current. If the number of turns of wire in the coil be increased or its radius decreased, the deflexion of the needle is increased. In all cases there is magnetic force exerted by the current and it remains to measure the force.

Now it is found that if a length l of wire be bent into the form of an arc of a circle of radius r, and if a current i is allowed to flow through the wire, the magnetic force at the centre of the arc is perpendicular to the plane of the arc and is proportional to il/r^2. The arc may be less than a complete circle or it may include one or more complete turns; if it include one turn exactly, then $l = 2\pi r$, and the force is proportional to $2\pi r i/r^2$ or $2\pi i/r$. If it include n turns then

$l = 2n\pi r$ and the force is then proportional to $2n\pi i/r$. In either of these cases lines of magnetic force are produced which stream through the coil and are linked with it.

Let the coil be placed in a vertical plane, and fix a horizontal sheet of card or glass so as to pass through its centre. Sprinkle the sheet with iron filings; on passing a current through the coil and tapping the sheet, the iron filings set along the lines of force as shewn in Fig. 142,

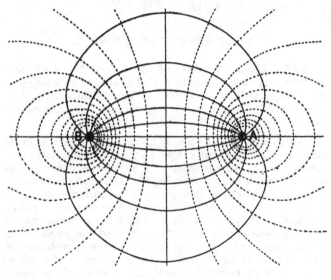

Fig. 142.

in which the dotted lines are lines of force, the strong lines equipotential surfaces. The lines could be mapped out on the sheet by means of a little compass-needle as described in Section 71.

In doing this it must be remembered that the magnet is also acted on by the earth's magnetism.

147. Unit of Current. If we assume the above law to hold we can deduce certain important consequences from it. The law stated that the force was proportional to this quantity. Let us assume the proportion to be one of equality, and let us further suppose that the current is measured in such a way that we may write

$$F = \text{Magnetic force at centre} = \frac{il}{r^2}.$$

Now let the length of the wire be 1 cm. and the radius of the circle 1 cm., then

$$l = 1, \qquad r = 1.$$

Hence $$F = i.$$

Thus the measure of the current is in this case equal to that of the magnetic force which it exerts at the centre of the coil. If this force be the unit force so that a magnetic pole of unit strength placed at the centre of the circle is acted on by a force of one Dyne, then we have $F = 1$, and therefore $i = 1$, or the current is the unit of current. This unit is known as the Electromagnetic Unit of Current.

DEFINITION. *The Electromagnetic* **Unit of Current** *is a current which, flowing in a wire one centimetre in length, bent into the form of an arc of a circle one centimetre in radius, produces unit magnetic force at the centre of the circle.*

The quantity of electricity which is conveyed by this current in unit time across each section of the conductor in which it is flowing is the electromagnetic unit quantity of electricity.

This electromagnetic unit is found by experiment to be very much greater than the electrostatic unit. Each electromagnetic unit contains 3×10^{10} electrostatic units.

It should be noted that the definition of the electrostatic unit is based on the supposition that the inductive capacity of air is unity, that of the electromagnetic unit depending as it does on the unit of magnetic force assumes the magnetic permeability of air to be unity, § 75.

148. Practical Unit of Current. The unit of current selected for practical purposes is, as we have said, the ampere. The ampere is defined in terms of the number of grammes of silver which are deposited by it, per second, from a solution of nitrate of silver. It has been found as the result of various experiments that the passage of the electromagnetic unit of current for 1 second causes the deposit of ·01118 gramme of silver. One-tenth of this current is taken as a practical unit, and is called an ampere.

DEFINITION. *A current which deposits per second ·001118 gramme of silver from a neutral solution of nitrate of silver in water is called one* **Ampere.**

The quantity of electricity conveyed by an ampere flowing for one second is, as we have already said, § 109, called a Coulomb. A **Coulomb** is one-tenth of the electromagnetic unit quantity of electricity.

149. Galvanometers. If a current traverse a wire bent into the form of a circle the direction of the magnetic force exerted near the centre of the circle is at right angles to its plane. If r cm. be the radius of the circle the force at the centre due to a current i is $2\pi i/r$. The force at a point near the centre is (if the circle is not too small) given approximately by this expression, and if a small magnet with poles of strength m be suspended at the centre each pole is acted on by a force $2m\pi i/r$. These forces act in opposite directions perpendicular to the plane of the coil. Each pole of the magnet will be acted on also by a force mH, where H is the strength of the earth's field, and the magnet will take up a position depending on the relation between these forces.

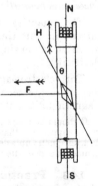

Fig. 143.

150. Tangent Galvanometer. Let us now suppose that the plane of the coil is north and south, and that the current above the magnet is running from south to north, below it from north to south. The force on the north pole of the magnet will be—applying the right-handed screw rule—from east to west and the north pole therefore will be deflected toward the west. Let F (Fig. 143) be the magnetic force toward the west, then the magnet is in equilibrium under two couples of which the forces are at right angles, the one F toward the west and the other H toward the north (§ 92). The tangent law holds, and if θ be the angle the magnet is deflected from the magnetic meridian we have

$$mF\cos\theta = mH\sin\theta$$

or
$$F = H\tan\theta.$$

Now we have already seen that if the current traverses a single turn of a circle

$$F = \frac{2\pi i}{r}.$$

Hence
$$\frac{2\pi i}{r} = H\tan\theta.$$

Therefore
$$i = \frac{Hr}{2\pi}\tan\theta.$$

If there are n turns in the coil traversed by the current

$$F = \frac{2n\pi i}{r}$$

and
$$i = \frac{Hr}{2n\pi}\tan\theta.$$

An instrument of this kind is called a tangent galvanometer.

Thus if we know the value of H we can by measuring r, the radius of the circle in which the current flows, and observing the value of θ, the deflexion of the magnet, calculate i the strength of the current, from the above formula.

The current so obtained will be in electromagnetic units. Since one ampere is one-tenth of such a unit, in order to find the current in amperes we must multiply by 10.

We thus have

$$i = \frac{10 \cdot Hr}{2n\pi}\tan\theta, \quad \text{amperes.}$$

Example. *Having given that the value of H is ·18 unit and that the radius of a coil of 10 turns is 5 cms., find the current in amperes which will cause a deflexion of 45°.*

In this case since $\tan 45 = 1$, we have

$$i = \frac{10 \times ·18 \times 5}{2 \times 10 \times \frac{22}{7}} = ·143 \text{ ampere.}$$

151. Reduction Factor. Sometimes it is convenient to denote the quantity $Hr/2n\pi$, or if we are working in amperes the quantity $10Hr/2n\pi$, by a single symbol k.

We then have $i = k \tan \theta$.

The quantity k is known as the reduction factor of the galvanometer; it depends on its construction and on the value of H at the spot where it is used but not on the deflexion.

In the case of a tangent galvanometer it may be defined as the quantity by which the tangent of the deflexion must be multiplied in order to give the current.

If we know the current required to give a certain deflexion we can find the reduction factor by dividing the current by the tangent of the deflexion.

Again, since $i = k \tan \theta$ and $\tan 45°$ is unity we see that if $\theta = 45°$ we have $i = k$. Thus the reduction factor is measured by the current required to produce a deflexion of $45°$.

Example. *A current of 10 amperes produces a deflexion of 60°, find the current which will produce a deflexion of 30°.*

Since $i = k \tan \theta$

we have $10 = k \tan 60$.

Hence $k = 10/\tan 60 = 10/\sqrt{3} = 5{\cdot}78$,

and the current required $= 5{\cdot}78 \tan 30 = 5{\cdot}78/\sqrt{3} = 3{\cdot}33$ amperes.

Or more simply, since the currents are proportional to the tangents of the deflexions, if i be the required current

$$\frac{i}{10} = \frac{\tan 30}{\tan 60} = \frac{1}{3}.$$

Hence $i = \frac{10}{3} = 3{\cdot}33$ amperes.

152. Galvanometer Constant. We have seen that the force exerted at the centre of the coil by a current i is $2n\pi i/r$, so that the force exerted by unit current is $2n\pi/r$. This quantity is called the galvanometer constant of the coil. It is often denoted by the symbol G, so that for a circular coil of n turns we have

$$G = \frac{2n\pi}{r}.$$

If the coil be not circular but have some other symmetrical shape, the magnetic force at its centre due to a unit current is still called the galvanometer constant, though its value cannot be so simply expressed as in the case of a circle.   In any case the force at the centre due to a current i is Gi in the direction of the axis, and if θ be the deflexion of the magnet we have

$$Gi = H \tan \theta$$

or
$$i = \frac{H}{G} \tan \theta.$$

Comparing this with the equation

$$i = k \tan \theta$$

we see that
$$k = H/G.$$

Thus the reduction factor is measured by the ratio of the strength of the earth's field to the galvanometer constant.

153. Sine Galvanometer.   In the tangent galvanometer the coil is placed in the magnetic meridian so that it is parallel to the needle when there is no current in the circuit.   The sine galvanometer is arranged so that the coil can be turned round a vertical axis through its centre.   When a current traverses the coil the needle is deflected and the coil is turned so as to follow the needle until its plane is again parallel to the needle in its deflected position. Thus the magnetic force due to the current in the coil at right angles to the magnet.   If we call this force Gi, and if θ be the deflexion as before

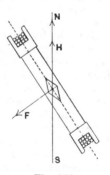

Fig. 144.

(Fig. 144), by taking moments round the centre of the magnet we obtain

$$Gi = H \sin \theta.$$

Therefore
$$i = \frac{H}{G} \sin \theta,$$

or writing
$$H/G = k$$
$$i = k \sin \theta.$$

For a circular coil the value of G is $2n\pi/r$. Thus the current is in this case measured by the sine of the deflexion and the instrument is a sine galvanometer.

It must be noted that in both these instruments it is assumed that the force acting on the magnet is determined by its value at the centre of the coil. This at best is only approximately true, and it is necessary that the magnet should be small compared with the radius of the coil.

Again, in the formula $F = 2\pi i/r$, r is the radius of a single turn in which the current flows, and the magnet hangs at its centre; when the coil has a number of turns, these are usually arranged in layers; the magnet clearly cannot be at the centre of all the coils in a given layer and the radii of the coils in different layers are different. Hence in the formula $F = 2n\pi i/r$, r has to be taken as the mean radius of the coil; if however the dimensions of a section of the coil are small compared with its radius, the mean radius can be determined with considerable accuracy.

154. Construction of a Tangent Galvanometer.

A tangent or sine galvanometer then usually consists of a circular coil of insulated wire wound
in layers in a suitable groove, as
shewn in Fig. 145. The coil is
carried on levelling screws and
mounted so that it can be rotated
round a vertical axis; a graduated
circle is attached to the coil, the
planes of the two being at right
angles, so that when the coil is
vertical the circle is horizontal; the
centres of the two coincide, and
at the centre of the circle is fixed
a pivot which carries a small magnetic needle. A light pointer is
attached to this and moves, as the
needle swings, over the graduated
circle; this pointer is usually fixed
at right angles to the axis of the magnet.

Fig. 145.

The circle is usually graduated so that when the axis of the magnet is in the plane of the coil the pointer reads zero on the circle. In using the instrument both ends of the pointer are read. This eliminates the error which might otherwise arise from a want of exact centering of the circle.

The magnet and pointer are protected from air currents by a transparent cover.

In a sine galvanometer a second horizontal circle is usually attached to the base of the instrument, and on this the angle through which the coil is turned can be read.

The ends of the wire are secured to binding screws on the base and the current is led into these. Sometimes a number of different circuits are wound in the same groove; the range of the instrument is thereby increased; for large currents a single turn may suffice to produce a measurable deflexion, for weaker currents a large number of turns may be necessary. Moreover a single turn of thick wire will have a low resistance which may be desirable for some purposes, while for others a coil of thin wire having a large number of turns and in consequence a considerable resistance may be required.

To use the instrument as a tangent galvanometer it is levelled so that the coil is vertical and the needle swings freely; then the coil is turned until it is parallel to the magnet, and in this case if the adjustments are correct both ends of the pointer read zero, or if the graduations run continuously round the circle, one end reads zero and the other $180°$. In this position the force exerted by the current will be perpendicular to that due to the earth. The current is then allowed to traverse the coil, thus deflecting the needle; as it comes to rest the instrument is gently tapped to reduce the effects of friction. Both ends of the needle are again read and the deflexion θ_1 is obtained by taking the mean of the deflexions given by the two ends; then if the reduction factor is known the current i is given by the formula

$$i = k \tan \theta_1.$$

If the current can be reversed through the coils, this is done and another deflexion θ_2 is found; the values of θ_1 and θ_2, if the adjustments are perfect, will be the same, but the small errors which may arise owing to (1) the pointer not being accurately perpendicular to the axis of the magnet, and (2) the zero of the scale being not quite correctly placed are eliminated by taking their mean.

To use the instrument as a sine galvanometer it is adjusted as before.

On allowing the current to traverse the coil the magnet is deflected, and the coil is turned round until the pointer again reads zero; in this position the magnet is again parallel to the coil and the force produced by the current is at right angles to the magnet.

The angle through which the coil has been turned is then read on the graduated circle at the base of the instrument. This gives one deflexion θ_1. By reversing the current and turning the coil in the opposite way we get a second reading θ_2. Then the current is given by

$$i = k \sin \tfrac{1}{2} (\theta_1 + \theta_2).$$

If there be no circle on the base of the instrument we may proceed as above, but after turning the coil carrying the current until the pointer reads zero the circuit is disconnected, when the magnet swings back into the meridian. The mean of the readings of the two ends of the pointer taken in this position will clearly give the angle the coil has been turned through.

In order to make the tangent galvanometer more sensitive a mirror is sometimes attached to the magnet; in this case the mirror and magnet are suspended by means of a fine silk fibre, or in some cases the fibre is of quartz. A lamp is arranged as described in Section 96 so that the light reflected from the mirror may fall on a scale, and the deflexion of the magnet measured by observing the position of the spot on the scale.

In some instruments the coil of wire is separated into two parts which are wound in two separate grooves. These are placed with their axes coincident, and the magnet hangs with its centre on the axis, midway between the two coils. If the distance between the coils be adjusted so as to be equal to the radius of either coil the magnetic field produced by the current in the neighbourhood of the magnet can be shewn to be far more uniform than when a single coil with the magnet at its centre is used. The theoretical conditions assumed for this instrument are more exactly fulfilled.

155. Sensitive Galvanometers. In a galvanometer the current is measured by the deflexion of a magnet which it produces.

An instrument in which a large deflexion is produced by a given current is more sensitive than one in which the deflexion is small.

Now we can increase the deflexion either by increasing the force which the current can exert or by decreasing the controlling force which maintains the magnet in its undisturbed position.

To increase the force due to the current we bring the coils of wire close to the magnet and increase the number of turns in the coil. As an increase in the number of turns means, for a given wire, an increase in the resistance, it does not follow that a large number of turns is always an advantage; the increase in resistance may, through reducing the current, reduce the force more than the increase in the number of turns increases it.

To reduce the control we may either (1) reduce the strength of the field in which the magnet hangs, or (2) adopt an astatic system. These methods may of course be combined.

(1) In the description of a tangent galvanometer given above, it has been assumed that the needle hangs in the magnetic field due to the Earth alone when the current is not on. This is not necessary; we can bring a bar magnet near to the instrument in such a position as to counteract much of the force due to earth; in this way the strength of the field in which the needle hangs can be much reduced and the deflexion due to a given current proportionately increased.

(2) In the astatic galvanometer two magnets are employed. These are rigidly attached with their axes parallel and their poles in opposite directions.

If the two magnets are exactly equal and their axes exactly parallel it is clear that when they are suspended in a uniform field, the couple on the one magnet will exactly balance that on the other, and the resultant will be zero. In practice this condition is never secured exactly, but by an arrangement of this kind the effect of the controlling force is greatly reduced.

Let the two magnets be suspended one above the other and suppose that, as shewn in Fig. 146, the galvanometer coils are placed so as to surround the lower magnet only, leaving the upper magnet above the coil.

The current in the coil will tend to turn both magnets

in the same direction though its effect on the upper magnet will be small compared with that on the lower.

Thus by this means the controlling force is considerably reduced and the sensitiveness increased.

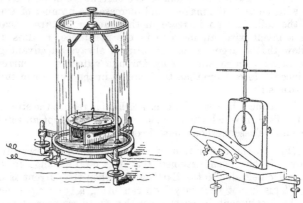

Fig. 146.   Fig. 147.

The accuracy with which the position of the galvanometer needle can be read is much increased by the use of a mirror as already described by which a beam of light is reflected on to a scale. Figs. 147 and 148 shew such instruments.

In Fig. 148 there are two magnets so that the system is astatic, and a second set of coils is added surrounding the upper magnet. The current passes in opposite directions round the two magnets, thus each magnet is subject to a considerable couple and these two couples tend to turn the system in the same direction.

The deflexion of the needle can be determined by observing the motion of the image reflected from the mirror.

Various arrangements are adopted in practice. Light from a lamp traverses a vertical slit and falls on the mirror. The scale is placed above the slit in a horizontal position; the mirror may be either concave or plane. If it be concave the slit and scale are placed at a distance from the mirror equal to its radius, and the scale is adjusted until the centre of the mirror is vertically above the slit and halfway between its centre and the scale. A real image of the slit is thus formed on the scale. If the mirror be flat a convex lens is used. This is sometimes placed close to the mirror so that both the incident and reflected beams traverse it. The lens should have a somewhat long focal length (say 1 metre) and its distance from the slit is equal to its focal length. The rays from the

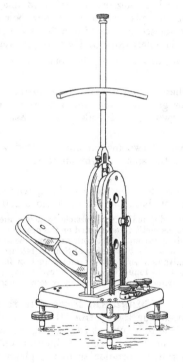

Fig. 148.

slit fall on the plane mirror as a parallel pencil and are re-flected as such back to the lens; after traversing it they form a real image of the slit on the scale.

If the light only traverses the lens once, it must be arranged as described in Section 96 to form a real image as far behind the mirror as the slit is in front. The light on its way to form this image is reflected by the mirror and the image is formed on the scale.

Instead of reflecting a beam of light on to a scale we may place a scale in front of the mirror, illuminate it suitably and then view with a telescope its reflected image. Another modification of this plan is to use a low power microscope in place of the telescope, attaching a fine scale just below the object glass of the microscope and viewing its reflected image.

Various forms of sensitive galvanometers are illustrated in Figs. 146—148.

Fig. 146 shews an ordinary astatic instrument, Fig. 147 a simple mirror galvanometer, and Fig. 148 a mirror galvanometer with two sets of coils and an astatic system of magnets.

Each of the last two instruments is fitted with a control magnet to bring the spot of light on to the scale and to vary the sensitiveness.

There are various other forms of a galvanometer and an account of some of these is given in Sections 220—222.

It should be noted that it is not possible to determine with accuracy by calculation the magnetic force exerted on the needle of a sensitive galvanometer by a given current. The coils are too close to the magnet and the conditions assumed in the theory of Section 146 are far from being satisfied. The galvanometer constant and reduction factor of such an instrument can only be found by a direct electrical experiment made for the purpose, employing a current measured by some other method.

156. Law of Magnetic force due to a Current.
We have deduced our theory from the assumption that the magnetic force due to a current i traversing an arc of a circle of length l cms. and radius r cms. at the centre of the circle is il/r^2 in a direction perpendicular to the plane of the circle. Various experiments have been devised to illustrate this law.

Thus, in Fig. 149, ABC represents a circular coil of given radius, say 5 cms., containing one turn. DEF is a second coil of twice the radius having two turns. The two coils are concentric, and their planes coincide. They are connected together so that the same current can circulate in opposite directions in the outer and inner coils respectively, and

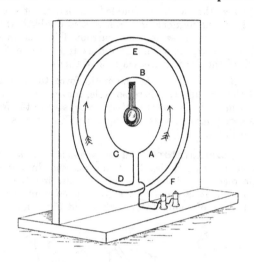

Fig. 149.

are placed with their plane in the magnetic meridian. A small magnet attached to a mirror is suspended at the common centre of the coils and a beam of light is reflected from the mirror on to a scale. The connexions are usually such that the current can be allowed to pass through either coil separately.

Pass the current through the inner coil; a deflexion of the spot of light is observed. Pass it through the outer coil, the spot is again deflected; if the current be not altered by the change of connexions it will be found that these two deflexions are equal.

Now pass the current through the two coils in series taking care that it circulates in opposite directions through the two. No deflexion will be observed. Thus the effect of the current in the single turn is balanced by that of the same current in two turns of twice the radius. We may shew in a similar way

that the effect of three turns of three times the radius is the
same as that of the single turn.

Now when the radius is doubled the length of each turn
is doubled but the number of turns is also doubled, thus the
length of the arc in which the current flows is increased
four-fold, or by 2^2, so when the radius is increased 3 times
the length of the arc is increased 9 or 3^2 times.

Hence if we suppose the magnetic force to vary as the
length of the arc, and this is reasonable, for each small element
of the arc must, it is clear, contribute equally to the force, it
must also vary inversely as the square of the radius. Thus
the force will vary as l/r^2.

157. Commutators. In experiments with electric
currents it is often desirable to reverse the direction of a
current in one part of a circuit without altering the battery
connexions. This is done by means of a commutator. One
convenient form of commutator is shewn in Fig. 150.

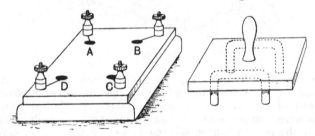

Fig. 150.

A, B, C, D are mercury cups arranged at the corners of a
square and connected to binding screws as shewn.

The battery wires are connected to A and C, the circuit in
which the current is to be reversed to B and D.

Two pieces of stout copper wire or rod are bent as
shewn in the figure and connected to an insulating handle.
The four ends of the rods fit into the mercury cups and
thus A can be connected to B and C to D, while by placing
the rods in the mercury cups with their lengths at right

angles to their original positions A and D are connected together, and also B and C.

Thus if A be connected to the positive pole of the battery, in the first position the current flows from A, then round the external circuit from B to D and thence to the battery from C, while in the second position the direction in the external circuit is from D to B.

Another form of commutator is shewn in Fig. 151.

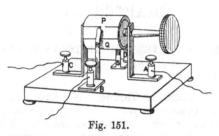

Fig. 151.

A cylindrical shaped piece of ebonite, which can be turned round its axis, carries two strips of brass P, Q at opposite extremities of a diameter of the cylinder. These are insulated from each other, but P is connected to a binding screw A by means of one of the supports in which the axis of the cylinder turns, while Q is connected to another screw C by means of the second support. These two supports are insulated from each other.

Two strips of stiff brass R and S connected to two other binding screws B and D press against the cylinder. If the cylinder were turned 90° in the right direction from the position shewn P would be in contact with R, Q with S. A current then entering at A passes to B, round the external circuit to D and then by the strips S and Q to C. If the cylinder be turned through 180° so that P is in contact with S, Q with R, the current entering at A passes through P, S, to D then round the external circuit from D to B and finally back by R and Q to C.

In addition to a commutator various other forms of keys and switches will be of service in electrical experiments; the mode of action of most of these will however be obvious from inspection.

CHAPTER XVI.

MEASUREMENT OF RESISTANCE AND
ELECTROMOTIVE FORCE.

158. Resistance Boxes. We have stated already that the resistance of a conductor is a physical property of the conductor which remains constant so long as the conductor retains the same physical conditions. A piece of wire therefore has a definite resistance and standards whose resistances are multiples or sub-multiples of the ohm can be constructed out of coils of insulated wire. It is found by experiment that the specific resistances of certain alloys such as Platinum-Silver, Platinoid, German Silver, "Manganin" (an alloy of Copper, Nickel and Manganese) are much higher than those of the pure metals and that their resistance changes less with temperature than does that of the pure metals. These substances therefore are chosen for resistance coils. Platinum-silver, an alloy of 67 parts of platinum and 33 of silver, is used for standard coils because of its permanence.

The coils are either wound on single bobbins, the ends of the wires being connected to binding screws, or are made up into resistance boxes.

In any case the wire is bent back on itself at its middle and wound double on the bobbin. This is to avoid the effect of self-induction[1] and to prevent direct magnetic action on any galvanometer-magnet, or similar instrument in the neighbourhood.

[1] See § 232.

The top of a resistance box is made of a non-conducting material and to this are attached a number of stout brass pieces as shewn at A, B, C, D in Fig. 152. A small space is left between each of these pieces of brass and the ends of these pieces are ground in such a way that a taper plug of brass can be inserted in them, and so put the brass blocks into electrical communication.

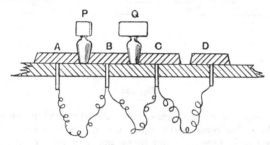

Fig. 152.

Binding screws are attached to the two end blocks. When the plugs are all inserted a current can pass from end to end through the blocks and plugs. The two ends of each coil are soldered to two consecutive blocks, so that when a plug such as P is removed, the current can pass from one block A to the next B only by traversing the coil. The resistance of the coil is thus introduced into the circuit between the binding screws.

The resistance of the brass blocks and plugs is so small as to be negligible for most purposes, provided the plugs fit properly, hence the total resistance between the binding screws is, in the case we have supposed, that of the coil connecting A and B.

The coils in a box are usually arranged thus:

1	2	2	5
10	10	20	50
100	100	200	500 units.

Thus with twelve coils as shewn any resistance between 0 and 1000 ohms can be inserted in the circuit.

In some boxes the coils are arranged in powers of 2 thus, $1, 2, 2^2, 2^3 \ldots$ units. Fewer coils are needed to make up a given resistance on this system than on the other.

Another arrangement is the dial box, the top of which is shewn in Fig. 153.

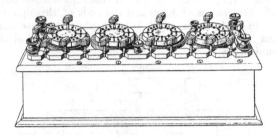

Fig. 153.

A series of eleven brass studs are arranged in an arc of a circle. The consecutive studs are connected by a series of 10 equal resistance coils and by means of a plug each can be connected in turn to a block at the centre of the circle. The connexions to the rest of the circuit are made at the centre of the circle and at the first stud.

When the plug connects the first stud to the centre block the current passes through it directly and the resistance of the arm is so small as to be negligible.

If the plug is placed in the second hole the current traverses the first resistance in passing through the box, if it be in the third hole two resistance coils are traversed by the current, and so on. Thus we can insert in the circuit any resistance between 0 and 10 units. A second dial, each coil of which is ten times that of the coils of the first dial, enables us to deal with resistances between 10 and 100 units, and so on.

159. Shunts. It may often happen that a current which is desired to measure produces too large a deflexion in a sensitive galvanometer. An arrangement whereby a definite fraction of the current, 1/10 or 1/100 of the whole, may be sent through the galvanometer is therefore convenient and this is secured by the use of a shunt.

A shunt is merely a resistance of suitable amount which is connected across the terminals of the galvanometer.

Thus let A, B, figure 154, be the terminals of the galvanometer and let R be its resistance, let A and B be connected through a resistance S. A current entering at

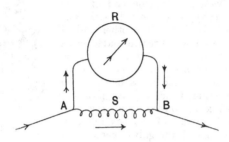

Fig. 154.

A has two paths to B and is divided in proportion to the conductivities of the paths. Let C be the whole current, C_g the part which passes through the galvanometer, C_s that which traverses the shunt.

Then $$C = C_s + C_g.$$

Also $C_g . R = C_s S$; for they both measure the potential difference between A and B.

Thus $$C_g = \frac{SC}{S+R},$$

$$C_s = \frac{RC}{S+R}.$$

These results of course follow at once from the formula of § 142.

Hence we have $$C = \frac{(S+R)C_g}{S}.$$

Thus if we know S and R and measure C_g we can find C.

In practice S is taken as some definite fraction of R. Thus if $S/R = 1/9$ we have

$$C = 10C_g;$$

or if $S/R = 1/99$, then
$$C = 100C_g.$$

A shunt box, constructed for a given galvanometer, usually contains three coils whose resist- ances are 1/9, 1/99, 1/999 of the resistance of the galvanometer. One end of each coil (Fig. 155) is connected with one terminal of the box. The other end of each goes to one of three insulated brass blocks attached to the top of the box. Each of these blocks can be connected by means of a plug to a fourth brass block which forms the other terminal. Suppose the plug placed so as to connect up the 1/99 coil. The terminals of the box are con- nected to those of the galvanometer and

Fig. 155.

99/100 of any current passes through the box, 1/100 through the galvanometer.

Another form of shunt box which can be used with different galvanometers is illustrated in figure 156. This was designed by Prof. Ayrton and Mr Mather.

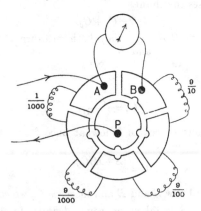

Fig. 156.

Let A, B be the terminals of a galvanometer of resistance R, let them be joined by a shunt of resistance S and let P be a point on the shunt circuit, the position of which can be varied. Let the resistance of AP be X, and let a current be introduced at A and withdrawn at P.

The fraction of this current which will traverse the galvanometer is $X/(R + S)$. That traversing the shunt is $(R + S - X)/(R + S)$.

By giving X the values S, $S/10$, $S/100$, $S/1000$ in turn, the fractions of the current traversing the galvanometer will be respectively

$$\frac{S}{R+S}, \quad \frac{1}{10}\frac{S}{R+S}, \quad \frac{1}{100}\frac{S}{R+S}, \quad \frac{1}{1000}\frac{S}{R+S},$$

assuming the current entering at A to remain constant.

Thus if the coil S be subdivided into four parts of $1/1000$, $9/1000$, $9/100$ and $9/10$ of the whole, and connected as shewn in Fig. 156 so that the block P can be connected by means of a plug to any of the divisions, the arrangement will serve as a shunt to any galvanometer.

160. Experiments on Electric Currents. We turn now to the methods of measuring the various electrical quantities, current, electromotive force, and resistance, with which we have been dealing.

Current and electromotive force are measured practically by some of the numerous forms of direct reading ammeters and voltmeters which are graduated to give the amperes or the volts, as the case may be, directly. Leaving these for the present[1] we will consider some experiments in electrical measurement which illustrate the various fundamental laws under discussion.

For the current measurements a tangent galvanometer is required.

An instrument with a coil about 10·5 centimetres in radius will be found useful. The groove should be wound with three separate coils, one of low resistance, containing 3 turns, a second of about 60 turns with a resistance

---

[1] See Sections 220—223.

considerably less than 1 ohm, the third with about 600 turns of thin wire having a resistance of 200 or 250 ohms. The terminals may be conveniently arranged so that the coils can be used either separately or in series.

We have seen that in a tangent galvanometer the current in C.G.S. units is given by the formula

$$i = k \tan \theta,$$

where k the reduction factor is equal to $Hr/2n\pi$, i is the current and θ the deflexion.

Putting in the values

$$H = \cdot18, \quad r = 10\cdot5, \quad n = 3, \quad \pi = 3\cdot14,$$

we find $k = \cdot1.$

Thus the values of k for the three separate coils are ·1, ·005, and ·0005, or measuring in amperes they are 1, ·05, and ·005; in other words, deflexions of 45° are caused respectively by currents of 1, ·05, and ·005 amperes.

For some of the experiments a mirror galvanometer, § 155, is required, while in some methods of comparing electromotive forces, and in verifying Ohm's law, a quadrant electrometer will be useful.

For reducing the results of the experiments a table of tangents will be required.

TABLE OF TANGENTS OF ANGLES FROM 0° TO 90°.

Angle	Tangent	Angle	Tangent	Angle	Tangent
0°	·000	13°	·231	26°	·488
1	·018	14	·249	27	·510
2	·035	15	·268	28	·532
3	·052	16	·287	29	·554
4	·070	17	·306	30	·577
5	·088	18	·325	31	·601
6	·105	19	·344	32	·625
7	·123	20	·364	33	·649
8	·141	21	·384	34	·675
9	·158	22	·404	35	·700
10	·176	23	·425	36	·727
11	·194	24	·445	37	·754
12	·213	25	·466	38	·781

Angle	Tangent	Angle	Tangent	Angle	Tangent
39°	·810	56°	1·48	73°	3·27
40	·839	57	1·54	74	3·49
41	·869	58	1·60	75	3·73
42	·900	59	1·66	76	4·01
43	·933	60	1·73	77	4·33
44	·966	61	1·80	78	4·71
45	1·00	62	1·88	79	5·15
46	1·04	63	1·96	80	5·67
47	1·07	64	2·05	81	6·31
48	1·11	65	2·15	82	7·12
49	1·15	66	2·25	83	8·14
50	1·19	67	2·36	84	9·51
51	1·24	68	2·48	85	11·4
52	1·28	69	2·61	86	14·3
53	1·33	70	2·75	87	19·1
54	1·38	71	2·90	88	28·6
55	1·43	72	3·08	89	57·3

161. Absolute measurement of a current.

EXPERIMENT 34.   *To measure absolutely*[1] *the current in a wire.*

For this purpose we need a galvanometer, the reduction factor of which we can determine by measurement; if the current be suitable we may employ the tangent galvanometer just described, using the coil of three turns if we can measure their diameter with sufficient accuracy. This may be done either with a pair of callipers[2], or by the aid of a steel tape which is stretched round the coil in contact with the wire.

Measure the diameter of the coil in centimetres and count the number of turns, then calculate the reduction factor from the formula $k = Hr/2n\pi$ using the value ·18 for H.

Set up the galvanometer as described in Section 154 and connect its terminals to two binding screws of a commutator or reversing key, Section 157. Connect the other two binding screws of the key to the battery or source of current, and observe the deflexion θ_1, reverse the key and again observe the deflexion θ_2. Then the current is given by the expression

$$c = k \tan \tfrac{1}{2} (\theta_1 + \theta_2).$$

[1] By the absolute measurement of a quantity is meant its determination in terms of the fundamental units of mass, length, and time.

[2] Glazebrook's *Mechanics*.

The current so found is in c.g.s. units. To obtain the value in amperes multiply by 10, for one c.g.s. unit contains ten amperes.

If the current is too small to be measured with the instrument described it may be desirable to use a mirror-galvanometer. A convenient instrument is formed by turning a small circular groove some 20 cm. in radius in a flat piece of wood, the groove being of such a size that a stout copper wire will just lie evenly in it to form the coil of the galvanometer. The wire forms as nearly as possible a complete circle, and its ends are carried away at right angles to the board, being kept as close together as is possible; if the wire is insulated the two ends are twisted together. This is to prevent direct action on the magnet. The board can stand in a vertical position. A hole of the form shewn in Fig. 157 is cut

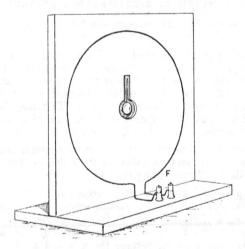

Fig. 157.

through the wood at the centre of the circle, and closed with two slips of glass, thus forming a small cell. In this cell a mirror is suspended; the mirror has some small magnets attached to its back. The board is placed in the magnetic meridian by the aid of a long magnet mounted as a compass-needle or in some other manner.

A lamp, slit and scale are arranged; using a lens, unless the mirror is concave, so as to produce an image of the slit on the scale. This image should be above the slit.

Instead of the lamp and slit an incandescent lamp may often be employed with advantage; a lamp with a straight filament is chosen, and a diaphragm arranged to cut off all the light except that from a straight bit of the filament.

The scale should lie in the magnetic meridian parallel therefore to the coil of the galvanometer; this is secured by turning it round until two points on it, equidistant from the slit, are equidistant from the magnet.

Measure the distance of the scale from the magnet; let it be a centimetres. Connect the two terminals of the galvanometer through the reversing key to a battery, and note the deflexion of the spot; let it be d_1 centimetres. Reverse the current and observe the deflexion in the opposite direction, d_2 centimetres. If the adjustments are perfect, d_1 will be equal to d_2, if the two be not exactly equal the mean d—equal to $\frac{1}{2}(d_1 + d_2)$—will be free from some small errors.

If θ be the deflexion of the magnet, since the reflected light is deflected through twice the angle through which the mirror is turned, we have

$$\tan 2\theta = \frac{d}{a},$$

and if d is small compared with a we have approximately

$$\tan \theta = \frac{1}{2}\frac{d}{a}.$$

But

$$i = \frac{Hr}{2n\pi}\tan\theta$$

$$= \frac{Hr}{2\pi} \times \frac{d}{2a}.$$

The value of r is found by measuring the diameter of the groove in which the wire lies with a finely divided scale or pair of callipers and correcting for the thickness of the wire. The value of H may be taken in England as ·180 units, hence, as all the quantities involved are known, the current is found.

Example. The radius of a single coil galvanometer is 20·5 cms.; the distance of the mirror from the scale is 100 cm., and a given battery when connected to the circuit produces a mean deflexion of 15·5 cm. Find the current.

If θ is the deflexion of the magnet

$$\tan \theta = \tfrac{1}{2}\,\frac{15\cdot5}{100} = \cdot0775.$$

The value of $Hr/2\pi$ is ·585.

Hence the value of the current is ·0454 c.g.s. units, or ·454 ampere.

162. Determination of Electro-chemical Equivalents.

EXPERIMENT 35. *To find the electro-chemical equivalent of copper.*

Connect up a galvanometer of known reduction factor with a reversing key, two or three cells of constant E.M.F. and a voltameter.

The voltameter consists of two copper plates supported in a piece of wood or ebonite so as to hang vertically and parallel to each other in a beaker containing a saturated solution of copper sulphate slightly acid.

Binding screws are attached to the plates so that the voltameter can be easily inserted in the circuit.

Clean the plates well with emery paper and weigh carefully one plate which is to serve as the kathode for the deposition of the copper. Let its mass be W grammes. Connect this plate to the zinc pole of the battery. Connect the other plate of the voltameter to one binding screw of the commutator. Connect the copper plate of the battery to another binding screw of the commutator. Connect the other screws of the commutator to the terminals of the galvanometer. Thus with the key in either position the current passes through the cell in the same direction, but is reversed in the galvanometer.

Make contact with the key and allow the current to flow for five minutes, reading the deflexions of the galvanometer each minute. Reverse the current in the galvanometer and allow it to flow for five minutes; then reverse again and after a third interval of five minutes, reverse a third time. Read the galvanometer at minute intervals throughout, and at the end of twenty minutes break the circuit. Remove the kathode plate from the solution. Wash it in distilled water, pour over it a little alcohol and then dry it quickly in a current of warm air—it should not be put in a flame. Weigh the plate again. Let its mass be W' grammes.

Then in twenty minutes the passage of the current has deposited $W' - W$ grammes of copper.

Determine the mean of the deflexions from the readings found during the passage of the current—these readings should not vary greatly. The mean is found by adding together the various deflexions and dividing the sum by the number of observations. Let it be θ and let k be the known reduction factor. Then the current is $k \tan \theta$. And this current has been flowing for 20×60 seconds. Thus the quantity of electricity which has passed is $20 \times 60 \times k \tan \theta$.

And the electro-chemical equivalent[1] being the ratio of the mass of copper deposited to the quantity of electricity which has passed is given by

$$(W' - W)/20 \times 60 \times k \tan \theta.$$

Example. The reduction factor of the given galvanometer is ·1, the mean deflexion is 46°, and the mass of copper deposited in 20 minutes is ·410 gramme. Find the electro-chemical equivalent of copper.

The value of tan 46° is 1·036.

Thus the electro-chemical equivalent is equal to

$$·410/120 \times 1·036$$

and this reduces to ·00330.

EXPERIMENT 36. *Having given the electro-chemical equivalent of copper, to find the reduction factor of a galvanometer.*

This is merely the converse of the last experiment. The arrangements and manipulation are the same. We are given however that the electro-chemical equivalent of copper is ·00329 gramme per C.G.S. unit of electricity.

---

[1] See Section 117.

Now the current flowing is the mass deposited per second divided by the electro-chemical equivalent.

Hence $\text{current} = \dfrac{W - W'}{20 \times 60 \times \cdot 00329}$.

But $\text{current} = k \tan \theta$.

Thus $k = \dfrac{W - W'}{20 \times 60 \times \cdot 00329 \times \tan \theta}$,

and the observations give us $W - W'$ and θ.

163. Observations on Ohm's Law.

According to Ohm's law if a constant current is traversing a wire the difference of potential between any two points is proportional to the resistance between them. Moreover for a uniform wire the resistance is proportional to the length. Hence in a uniform wire carrying a constant current the potential difference between two points is proportional to the distance between them. Hence if AB, Fig. 158, be a straight wire

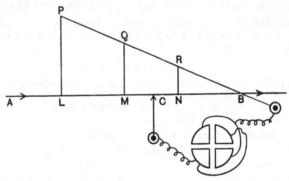

Fig. 158.

carrying a current, L, M, N a number of points on the wire, and if lines LP, MQ, NR, etc. be drawn from these points at right angles to AB to represent the potential differences between L and B, M and B, N and B, etc., then it is clear that since $PL/BL = QM/BM = RN/BN$, the points P, Q, R lie on a straight line which passes through B.

EXPERIMENT 37. *To verify Ohm's Law.*

The experiment may be carried out thus :

Some form of electrometer is needed to measure the volts.

A fine wire of German-silver or some other material of large specific resistance is stretched along a divided scale, being attached to two terminal screws A and B. A slider moving along the scale carries another screw C and a contact piece which allows of contact being made between a wire attached to C and any desired point on the stretched wire.

The screw B is connected to one pole of a constant battery—a small number of Daniell cells or one or two storage cells—and to one pair of quadrants of a quadrant electrometer. The screw A is connected to the other pole of the battery, while C is in connexion with the other pair of quadrants of the electrometer.

The readings of the electrometer then give the difference of potential between B and C. Commence from B and make contact with the slider in succession at a number of points along the wire noting the electrometer deflexion at each point.

Then plot a curve, taking for abscissæ the distances from B of the successive points of contact, and for ordinates the electrometer readings. The curve will be found to be a straight line which passes through B. Thus the law is verified that when the current is constant the difference of potential between two points is proportional to the resistance.

Again we have to shew that for a given resistance the electromotive force is proportional to the current.

To do this we may employ the same arrangements as above, but we include in the circuit between the wire AB and the battery, a galvanometer for measuring the current—an ammeter—and a resistance box.

Measure with the electrometer the volts between A and some fixed point L of the wire; let them be E_1. Measure also the current in the main circuit; let it be i_1. Put some resistance into the circuit by taking plugs out of the box, and thus reduce the current to i_2. Measure the volts; let them be E_2. Continue thus for various values of the currents. Then it will be found that we have

$$\frac{E_1}{i_1} = \frac{E_2}{i_2} = \ldots,$$

or if we plot a curve, taking the currents as abscissæ and the volts as ordinates, the curve will be a straight line.

If the galvanometer used to measure the current be a tangent instrument, and if the deflexions be θ_1, θ_2, ... then

$$\frac{i_1}{\tan \theta_1} = \frac{i_2}{\tan \theta_2} = \ldots = k,$$

where k is the reduction factor.

Hence the formulæ become

$$\frac{E_1}{\tan \theta_1} = \frac{E_2}{\tan \theta_2} = \ldots.$$

And we take as abscissæ of the curve the values of $\tan \theta$.

These two experiments completely verify Ohm's law.

164. Voltmeters. A voltmeter is a galvanometer which is arranged to measure volts.

Suppose L, M are two points on a wire carrying a current; there is a certain difference of potential, E volts say, between the two. If L and M be now connected by wires to a galvanometer, the potential difference is reduced and some of the current flows through the galvanometer; if however the resistance of this circuit is very great compared with that of LM, the potential difference between L and M will be altered by a very small amount, we may still take it as E volts, and the very small current through the galvanometer will be proportional to E. By using a sensitive galvanometer we may measure this current with accuracy, and hence, if we know the resistance in the galvanometer circuit, determine by the use of Ohm's law the potential difference between L and M.

Thus for example if the resistance of the galvanometer circuit be 1000 ohms, and if a current of ·001 ampere is observed, the potential difference is ·001 × 1000 or 1 volt.

A galvanometer so arranged is called a voltmeter.

165. Experiments on Batteries.

EXPERIMENT 38.   *To compare the electromotive forces of different batteries.*

(1) *With the tangent galvanometer.*   Adjust the galvanometer as already described and connect each battery in turn to the long coil, using the reversing key. The resistance of most ordinary batteries will be small compared with the 250 ohms of the long coil so that we may assume without great error that the various batteries in turn are working through the same resistance. Thus the currents produced are proportional[1] to the electromotive forces and the currents are proportional to the tangents of the deflexions; hence if E_1, E_2 be the E.M.F.'s of two batteries, θ_1, θ_2 the deflexions, then

$$E_1 = Ri_1 = Rk \tan \theta_1,$$
$$E_2 = Ri_2 = Rk \tan \theta_2,$$

hence

$$\frac{E_1}{E_2} = \frac{\tan \theta_1}{\tan \theta_2}.$$

Example. *When a Daniell cell is connected to the galvanometer the mean deflexion is 39°, when a Leclanché cell is connected it is 47·30°. Compare the electromotive forces.*

We have

$$\frac{\text{E.M.F. Leclanché}}{\text{E.M.F. Daniell}} = \frac{\tan 47\cdot30}{\tan 39} = \frac{1\cdot09}{\cdot81} = 1\cdot34.$$

If we take the E.M.F. of the Daniell as 1·07 volts, we find for that of the Leclanché 1·43 volts.

There are various ways in which we can allow for the effect of the battery resistance.

Thus arrange the two cells in series joining the negative pole of one to the positive pole of the second, and then connecting to the galvanometer the positive of the first and the negative of the second. The E.M.F. will be $E_1 + E_2$, the sum of the E.M.F.'s of the two; the resistance will be

---

[1] In reality $i = E/(B + R)$, where B is the battery resistance, R that of the rest of the circuit. If B is small compared with R we may write without great error $i = E/R$, and R is the same for all the batteries.

$R + B_1 + B_2$, R being the external resistance, B_1, B_2 the battery resistances. Observe the deflexion reversing the current as usual, let the mean deflexion be θ.

Now join the two negative poles together and connect the two positives to the galvanometer. The cells oppose each other, thus the E.M.F. is $E_1 - E_2$, but the resistance is $R + B_1 + B_2$ as before. Let the mean deflexion be θ'. Then the resistances being the same the currents are proportional to the electromotive forces. Hence

$$\frac{E_1 + E_2}{E_1 - E_2} = \frac{\tan \theta}{\tan \theta'}.$$

Example. *With the same two batteries as in the previous example it is found that* $\theta = 62°$, $\theta' = 15°\ 30'$.

Hence
$$\frac{E_1 + E_2}{E_1 - E_2} = \frac{\tan 62°}{\tan 15°\ 30'} = \frac{1\cdot 88}{\cdot 28}.$$

From this will be found that
$$\frac{E_1}{E_2} = \frac{2\cdot 16}{1\cdot 60} = 1\cdot 35.$$

Or if
$$E_2 = 1\cdot 07 \text{ volts},$$
$$E_1 = 1\cdot 44 \text{ volts}.$$

The difficulty about the method is that if E_1 and E_2 are nearly equal then θ' is very small and a small error in the angle makes a considerable error in the result.

(2) *With a mirror galvanometer.* The theory is the same but (a) for small deflexions we may assume the current to be proportional to the deflexion, and (β) a large resistance will of necessity be inserted in the circuit to render the current sufficiently small to be measured on a sensitive instrument, and this large resistance will diminish the error due to the omission of the battery resistance.

Adjust the galvanometer and bring the spot of light to the centre of the scale. Connect up in series a resistance box, the battery and the galvanometer.

Take sufficient resistance out of the box to give some convenient deflexion δ_1. If a Daniell cell be used as the standard, a deflexion of 107 scale divisions will be a suitable value for δ_1. Replace the Daniell by the battery to be tested and read the deflexion δ_2. Then if we may neglect the battery resistance, and when several thousand ohms resistance are included in the circuit this may usually be done without sensible error, we have

$$\frac{E_2}{E_1} = \frac{\delta_2}{\delta_1}.$$

If we have made $\delta_1 = 107$ for the Daniell for which $E_1 = 1 \cdot 07$ volts, then clearly $E_2 = \delta_2/100$ volts.

This method is sometimes varied by adjusting the resistance in the circuit until the deflexions due to the two cells are equal.

We know then that the currents are equal, and this too without making any assumption as to the relation between current and deflexion. We have then that the electromotive forces are proportional to the whole resistance in circuit in the two cases.

Let G be the galvanometer resistance, B_1, B_2 the two battery resistances. Connect the one battery and insert a resistance R_1 to give some convenient deflexion. Replace this battery by the second and insert a resistance R_2 to give the same deflexion, then the resistances in the two cases are $B_1 + G + R_1$ and $B_2 + G + R_2$.

Hence $$\frac{E_1}{E_2} = \frac{B_1 + G + R_1}{B_2 + G + R_2}.$$

In practice it very often happens that B_1, B_2 and G are all very small compared with R_1 and R_2, and in this case

$$\frac{E_1}{E_2} = \frac{R_1}{R_2}.$$

(3) *By the Potentiometer method.* In Fig. 159 AB represents a long fine wire 1 or 2 metres in length of considerable resistance. The end A is connected to the positive pole and the end B to the negative of a constant battery —a storage cell is suitable. Thus a current is running from A to B and there is a steady drop of potential along the wire. The positive pole of one of the batteries to be tested is connected to A, the negative pole is connected through a sensitive galvanometer G to a sliding contact piece P with which contact can be made anywhere on the wire.

Now the main current produces a potential difference between A and P and in consequence a current tends to flow through the battery under test and the galvanometer between

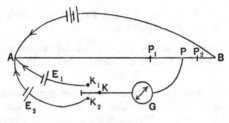

Fig. 159.

A and P. But the electromotive force E_1 of this battery tends to drive a current round the same circuit in the opposite direction. If E_1 is greater than the drop in volts between A and P due to the main current, the current in the galvanometer will be from P to A; if E_1 is less than the drop in volts the current will be from A to P.

By shifting the slider a position can be found for P for which there is no current through the galvanometer. When this is the case the E.M.F. E_1 must be equal to the drop in volts between A and P, and so long as the main current is constant this drop in volts is proportional to the resistance and hence to the distance AP.

Thus if P_1 be the position so found we see that, if the main current is constant, E_1 is proportional to AP_1.

The first battery is now removed and its place taken by the second, of E.M.F. E_2; if another position P_2 be found for which there is no current, then E_2 is proportional to AP_2.

Hence
$$\frac{E_1}{E_2} = \frac{AP_1}{AP_2}.$$

In practice the positive poles of the two batteries to be compared are permanently connected to A. Their negative poles are connected to two of the terminals K_1, K_2 of a two-way switch. The common terminal of the switch is connected to the galvanometer and the galvanometer to the slide. With the switch in one position K is connected to K_1, the first battery is in position and the point P_1 is found. Then K is switched over to K_2, the second battery is connected and P_2 is found.

To eliminate any small change in the main current the key is again transferred to K_1 and another position found for P_1; if the two positions do not differ much the mean of the two corresponding lengths is taken as the measure of E_1, while E_2 is measured by AP_2.

This method is known as Latimer Clark's Potentiometer method of comparing electromotive forces.

As the standard a Latimer Clark or a Weston cell is generally used. Since the E.M.F. of a Clark cell at 15° C. is 1·434 volts, it is convenient to arrange if possible that the length AP for the standard should be 1·434 metres or 1434 millimetres. When this is done the reading in metres corresponding to any other battery gives its E.M.F. in volts.

This result is attained by inserting a resistance between B and the negative pole of the main battery. When the Clark is in circuit the slider P is set to the required point 1434 mm. from A, and the main current adjusted by means of this resistance until the galvanometer is not deflected. When this is the case the drop of potential for each metre of the wire is 1 volt.

If the temperature be not 15° C. the reading for the Clark cell must be set to represent its E.M.F. at the time of the observation.

In practice the whole or a part of the wire can often be conveniently replaced by a series of resistance coils, each coil being equal in resistance to say 100 divisions of the wire. The wire need then only be 100 divisions long. Thus for the Clark cell we should need 14 of the coils and 34 divisions of the wire; for a cell of voltage 1·080 volts we need 10 coils and 80 divisions of the wire.

166. The Potentiometer. (i) *Measurement of current.*
The method can be employed to measure a current in the
following manner :

The current to be measured is passed through a known
resistance R, of sufficient size to carry the current. Thus
a fall of potential Ri is set up between the ends of the
resistance, i being the current.

The end of this resistance at which the current enters is
connected with the terminal A of the wire (Fig. 160), the other

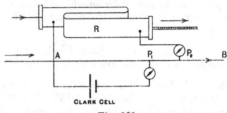

Fig. 160.

end being connected with the screw K_2 of the switch shewn
in Fig. 159[1]. Thus the resistance carrying the current takes
the place of the battery whose E.M.F. is to be found in terms
of the Clark or other standard cell, and the fall of potential
Ri is measured in the same manner as the E.M.F. Hence
if P_1, P_2 be as before the positions of the sliding contact
(1) with the standard in circuit, (2) with the resistance and
current connected, we have

$$\frac{Ri}{E_1} = \frac{AP_2}{AP_1}.$$

Hence $$i = \frac{E_1}{R} \cdot \frac{AP_2}{AP_1}.$$

(ii) *Measurement of resistance.* Two resistances R, S
can also be compared by this means.

A current i is passed through the two resistances in series
(Fig. 161): a second two-way switch has its terminal connected to the terminal A. The positive ends of the two

[1] In Figs. 160 and 161 the keys are not shewn. The figures indicate
the condition when the balance is reached. One galvanometer only is
used, being connected as in Fig. 159 to each circuit in turn.

resistances R, S are connected to the other terminals of the switch; the negative ends of R and S being connected to K_1 and K_2.

Thus the switches can be arranged so that in one position R is connected to A and through the galvanometer to the slider; in the other R is out of circuit and S takes its place.

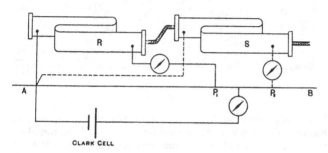

CLARK CELL

Fig. 161.

Now if i be the common current in R and S, in the one position the E.M.F. in the slider circuit is Ri, in the other it is Si. Hence if P_1, P_2 are the positions of the slider

$$\frac{Ri}{Si} = \frac{AP_1}{AP_2}.$$

Thus
$$R = S\frac{AP_1}{AP_2}.$$

In many cases the wire and slider may be usefully replaced by two resistance boxes. See Glazebrook and Shaw, *Practical Physics*, Section W.

In Fig. 162 we have a figure of the instrument as now arranged by Messrs Crompton. The potentiometer coils are shewn to the left, and the rheostat for adjusting the main current to the right; the centre dial takes the place of the keys K_1, K_2 of the previous description; by it various circuits in turn can be connected to the galvanometer.

167. The Potentiometer. Further applications.

We may use the apparatus described in the previous sections

to verify the laws as to the effects of connecting batteries in series and in parallel. Thus measure by one or other of the above methods the electromotive forces of two or more cells. Then arrange the cells in series, the positive pole of one cell

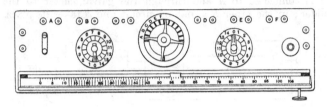

Fig. 162.

being connected to the negative of the preceding one, and measure the E.M.F. of the combination.   It will be found that this is the sum of the electromotive forces of the separate cells.

Now arrange the cells in parallel so that all the positive are connected together and all the negative poles, and measure the E.M.F. again ; if the cells all have the same E.M.F. the result will be found to be equal to the E.M.F. of one cell ; if the cells differ in E.M.F. the result will be intermediate between those of the given cells ; it will depend on the resistances and cannot be calculated unless these are known.

168. Batteries in series and in parallel. Again connect each cell up in turn first to the short coil then to the long coil of the tangent galvanometer and note the deflexion in each case.

Arrange the cells in series and connect them to the long coil. The current flowing will be greater than that due to each cell singly, in fact if θ_1, θ_2, θ_3 be the deflexions due to the single cells, θ that due to the cells in series, then unless one of the cells has an abnormally high resistance, it will be found that we have approximately

$$\tan \theta = \tan \theta_1 + \tan \theta_2 + \ldots$$

For if E_1, E_2, ... be the electromotive forces, B_1, B_2 the resistances of the batteries, i_1, i_2, ... the currents, R the galvanometer resistance, and i the current when the cells are in series, we have, taking the case of two cells,

$$i_1 = \frac{E_1}{R + B_1}, \qquad i_2 = \frac{E_2}{R + B_2},$$

$$i = \frac{E_1 + E_2}{R + B_1 + B_2}.$$

Now with the long coil of the galvanometer R is large compared with B_1 or B_2. Hence we may write approximately

$$i_1 = \frac{E_1}{R}, \qquad i_2 = \frac{E_2}{R},$$

$$i = \frac{E_1 + E_2}{R} = i_1 + i_2.$$

And since the currents are proportional to the tangents of the deflexion

$$\tan \theta = \tan \theta_1 + \tan \theta_2,$$

or, in words, when the external resistance is large compared with the battery resistance the current produced by a number of cells in series is very nearly the sum of the currents due to the individual cells.

For working a long telegraph line a high voltage, obtained by using a large number of cells, is required. We may compare this with the problem of forcing water through a tube of narrow bore. A pump capable of exerting great pressure is needed, the resistance caused by the water having to pass through the valves of the pump is small compared with that which arises from the narrow bore of the pipe. Nothing much is gained by increasing the passages in the pump. The pressure which it can exert determines the flow.

Now connect the cells still arranged in series to the short coil. The current is not much greater than that given by one cell alone. The electromotive force it is true has been increased, but the resistance being chiefly due to the battery —the resistance of the short thick coil is very small compared with the resistance of the cell—the resistance of the circuit has been increased with the increase in E.M.F. and there is no gain in current,

Taking the cells, n in number, as all equal, the E.M.F. is nE and the resistance nB. Thus we have

$$i = \frac{nE}{nB + R},$$

and since R is small compared with B we may neglect it in the denominator and hence

$$i = nE/nB = E/B$$

= current due to one cell through the short coil.

Now arrange the cells in parallel and connect them to the long coil; the current does not differ much from that due to a single cell. The E.M.F.—assuming the cells alike—is the same as that of each cell, the battery resistance is reduced by the combination, but in any case the battery resistance is a small fraction of the total resistance, thus the total resistance is not appreciably altered and the current is nearly the same as that due to one cell.

Connect the cells, still arranged in parallel, to the short coil. The current is considerably greater than that given by a single cell; for the battery resistance is in this case the main portion of the total resistance and the battery resistance is reduced by connecting the cells in parallel.

To put the result in symbols, if there be n cells all alike connected in parallel, the E.M.F. is E and the resistance B/n.

Thus the current is given by

$$i = \frac{E}{R + B/n}.$$

For the long coil B/n is small compared with R, and the current is approximately E/R, the same as that due to one cell through the long coil; for the short coil R is small compared with B/n so that the current is nE/B or n times that due to one cell working through the short coil.

169. Comparison of Resistances. Ohm's law is the basis of various methods of comparing resistances.

EXPERIMENT 39. *To compare two resistances by the use of a tangent galvanometer and resistance box or coil of known resistance.*

The coil of the galvanometer which has about 60 turns and a resistance of less than 1 ohm may conveniently be used.

Connect in series a constant cell, the galvanometer, adjusted as already described, a key, and the unknown resistance R. Let B be the battery resistance, G the galvanometer resistance, and let θ_1 be the deflexion. Then the total resistance in circuit is $B + G + R$.

Replace the unknown resistance by the standard S and let θ be the deflexion. The total resistance is $B + G + S$.

The currents are in the ratio of $\tan \theta_1$ to $\tan \theta$, but the E.M.F. being the same in the two cases the currents are inversely proportional to the resistances. Hence they are also in the ratio of $B + G + S$ to $B + G + R$.

Thus
$$\frac{B + G + R}{B + G + S} = \frac{\tan \theta}{\tan \theta_1}.$$

Now S is known; thus if B and G are known R can be found. In many cases we are sure that B and G are so small that we may neglect them; when this can be done
$$\frac{R}{S} = \frac{\tan \theta}{\tan \theta_1}.$$

Hence
$$R = S \cdot \frac{\tan \theta}{\tan \theta_1}.$$

If a resistance box is used so that S is adjustable we can proceed thus.

Take plugs out of the box until the deflexion in the second case is equal to that in the first, then the currents are equal and the electromotive force is the same. Hence the resistances must be the same.

Hence $B + G + R = B + G + S.$

Hence $R = S.$

This assumes of course that the box is such that it is possible to find a resistance S in it equal to the given resistance. Clearly this is not always possible, but if the box is divided to ohms we can find two resistances S ohms and $S + 1$ ohms between which R must lie. This will often be sufficiently accurate for our purposes.

To carry out the experiment the apparatus may be arranged as in Fig. 163 in which K, K_1, K_2 is a two-way switch. With the switch connecting K and K_1 the current passes

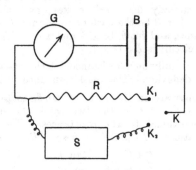

Fig. 163.

through R, the standard S is out of circuit; with K and K_2 in connexion the current passes through S, the unknown coil is out of circuit.

The formula given above assumes that a tangent galvanometer is used; this may in practice be replaced with advantage by a direct reading ammeter.

EXPERIMENT 40. *To examine the manner in which the resistance of a wire depends on its length, sectional area, and material.*

For this purpose there are given two coils, each marked A, of German-silver or some other material of high resistance of the same length and cross section, a third coil B of the same material and length as A but with double the sectional area, and a fourth coil C of iron of the same length and sectional area as A.

First place A in circuit and observe the deflexion θ_1, then introduce the second coil A in series with the first, let the deflexion be θ_2. Replace the two coils by B, let the deflexion be θ_3; finally substitute C for B and let the deflexion be θ_4.

Look out the tangents of the various angles. Then if we

assume the resistances of the battery and galvanometer to be small, and write A, B, C for the resistances of the three coils, we have seen that the currents—measured by the tangents of the deflexions—are inversely as the resistances,

$$\frac{A \text{ and } A \text{ in series}}{A} = \frac{\tan \theta_1}{\tan \theta_2},$$

and we shall find that $\tan \theta_1 = 2 \tan \theta_2$. Thus the resistance of two equal coils in series is double that of either coil.

Again
$$\frac{B}{A} = \frac{\tan \theta_1}{\tan \theta_3},$$

and we shall find that $\tan \theta_3 = 2 \tan \theta_1$.

Hence $B = \frac{1}{2}A.$

Thus by doubling the cross section we halve the resistance; the resistance of a coil of given length is inversely proportional to the area of its cross section.

Again C and A are of the same length and cross section but of different materials, the ratio $\tan \theta_1 / \tan \theta_4$ measures the ratio of the specific resistances of iron and German-silver.

By arranging coils in series or in parallel and measuring their resistances we can verify the laws given in Sections 139, 140.

Thus let the deflexions be as below:

With a standard coil S in circuit θ;

With A in circuit θ_1, with C in circuit θ_4;

With A and C in series θ_5;

With A and C in parallel θ_6.

Hence
$$A = S \frac{\tan \theta}{\tan \theta_1},$$

$$C = S \frac{\tan \theta}{\tan \theta_4},$$

$$A \text{ and } C \text{ in series} = S \frac{\tan \theta}{\tan \theta_5},$$

$$A \text{ and } C \text{ in parallel} = S \frac{\tan \theta}{\tan \theta_6}.$$

On evaluating these expressions we find that the resistance of A and C in series is equal to $A + C$, while the resistance R of A and C in parallel satisfies the relation

$$\frac{1}{R} = \frac{1}{A} + \frac{1}{C},$$

so that $\qquad\qquad R = AC/(A + C).$

In most of the above experiments it has been supposed that the resistances of the battery and galvanometer are negligible. If this is not the case experiments[1] can be arranged to find them, but these are complicated and in any case the methods given in the following section are free from this and other defects of the methods just described which assume (1) that the E.M.F. of the battery remains constant, and (2) that the current can be read with sufficient accuracy on the tangent galvanometer or ammeter, if a direct reading ammeter be employed. Frequently neither of these conditions is satisfied.

170. Wheatstone's Bridge. Consider an arrangement in which (Fig. 164) two circuits ACB, ADB are open to a current flowing into the circuit at A and out at B.

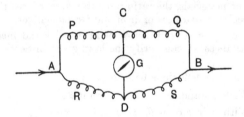

Fig. 164.

The potential at A is higher than that at B and falls as we pass along either circuit from its value at A to its value at B. If we take a point C on the one circuit we can find a point D on the other which has the same potential as C. To do this connect a wire to the circuit at B and to one terminal of a galvanometer G. Attach a second wire to the other terminal of the galvanometer, and make contact with the loose end at various points on the circuit ADB. If the point of contact be near A, the galvanometer indicates

[1] Glazebrook and Shaw, *Practical Physics.*

a current in one direction, if the point of contact be near B, the current in the galvanometer is in the opposite direction; a position can be found for which there is no current in the galvanometer. Let this position be D. Then it is clear that C and D are at the same potential.

Thus

Fall of potential from A to C = Fall of potential from A to D,

and

Fall of potential from C to B = Fall of potential from D to B.

Now when a given current is flowing in a wire the fall of potential between any two points in the wire is proportional to the resistance between these points.

Thus

$$\frac{\text{resistance from } A \text{ to } C}{\text{resistance from } C \text{ to } B} = \frac{\text{fall from } A \text{ to } C}{\text{fall from } C \text{ to } B}$$

$$= \frac{\text{fall from } A \text{ to } D}{\text{fall from } D \text{ to } B} = \frac{\text{resistance from } A \text{ to } D}{\text{resistance from } D \text{ to } B}.$$

Thus let P ohms = resistance of AC,

                Q ohms =   ,,   ,, CB,

                R ohms =   ,,   ,, AD,

                S ohms =   ,,   ,, DB.

Then we have arrived at the result that when there is no current between C and D we must have

$$\frac{P}{Q} = \frac{R}{S}.$$

If then three of the resistances P, Q and S are known we can find the fourth R, while clearly it is not necessary to know the actual values of P and Q provided the ratio P/Q is known.

We may arrive at the same result graphically thus:

Let AC, CB, Fig. 165, represent two resistances P, Q respectively. Let AL at right angles to AB be the E.M.F. in the circuit ACB. Join LB, and draw CM parallel to AL to meet LB in M and MN parallel to AB to meet AL in N.

Then clearly CM measures the potential difference between C and B.

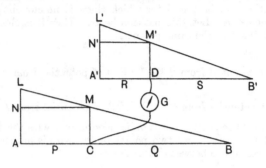

Fig. 165.

Again let $A'D$ represent R and DB' S, D being the point on ADB which is at the same potential as C, and construct a similar figure for the circuit ADB. It is clear from the figure that there is such a point, for A and A' are at the same potential as also are B and B'. Then since the E.M.F. between A and B is the same along either path we have $AL = A'L'$, and since the E.M.F. between C and B is the same as that between D and B we have $CM = DM'$.

Hence $$LN = L'N'.$$

Thus we must have

$$\frac{P}{Q} = \frac{AC}{CB} = \frac{MN}{CB} = \frac{LN}{CM} = \frac{L'N'}{DM'} = \frac{M'N'}{CB} = \frac{A'D}{CB} = \frac{R}{S},$$

or as before $$\frac{P}{Q} = \frac{R}{S},$$

whence $$R = S \cdot \frac{P}{Q}.$$

When this relation is satisfied so that an E.M.F. in the branch AB produces no current in the branch CD, these two conductors are said to be conjugate to each other.

The apparatus for measuring resistance by this method takes various forms.

EXPERIMENT 41. *To measure a resistance by Wheatstone's bridge method with a wire bridge.*

A fine uniform wire is stretched along a metre scale (Fig. 166). This constitutes the conductor ACB, a sliding

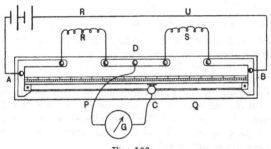

Fig. 166.

piece moving along the scale permits of contact being made at any point C of the wire. The ends A, B of the wire are soldered to two thick plates of copper each of which carries two binding screws. A third thick plate of copper fixed between the other two has three binding screws attached to it. This constitutes the point D of Fig. 165. The wire whose resistance it is required to measure is attached between A and D. Between D and B is placed a known resistance S. The wires from a battery are joined to the apparatus at A and B; the battery circuit should also contain a key. Wires from a galvanometer, usually an astatic or other sensitive instrument, are joined to C and D.

The battery circuit is closed, and the slider is moved along the scale making contact at various points until a position is found for which the galvanometer needle is not deflected.

When this result is attained we know that

$$R/S = P/Q.$$

Read the position of C on the scale and thus measure the lengths AC and CB, let them be a cm. and b cm. respectively.

Then since ACB is a uniform wire the resistance of any portion of it is proportional to its length.

Thus $P/Q = a/b.$

Hence $R/S = a/b,$

or $R = S\dfrac{a}{b}.$

If then S is known R can be found.

The resistance S may either be a single coil of wire having a known value, or more conveniently a resistance box out of which suitable plugs have been taken; the measurements, it can be shewn, will be most accurate when a is as nearly as possible equal to b. This condition can be approximately secured if S is adjustable. A value is taken for S and then R is found. The resistance S is then altered so as to be nearly equal to this value of R and then the experiment is repeated giving a more accurate value for R.

EXPERIMENT 42.   *To find the resistance of a coil by Wheatstone's bridge method using a Post Office box of coils.*

In this method the three resistances P, Q, S are taken from a resistance box (Fig. 167). Each of the arms AC, CB

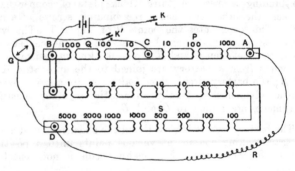

Fig. 167.

of the bridge usually contains three coils whose resistances are 10, 100, and 1000 ohms respectively. Thus by taking a single plug out of AC, P may have any of the values 10, 100 or 1000 ohms, while by taking a single plug from CB, Q may be either 10, 100 or 1000.

These coils are spoken of as the ratio arms of the bridge and the value of P/Q may be either ·01, ·1, 1, 10 or 100.

The arm S or BD contains a number of coils by means of which any resistance between 1 and 10,000 ohms may be unplugged. It is important to take care before beginning an experiment that all the plugs are firmly in their places.

Double binding screws are attached to the box at A and D, and single screws at B and C; the box is generally so arranged that by removing a copper strip the connexion at B between S and Q may be broken.

A sensitive reflecting galvanometer is connected with a key in circuit between B and D; while the battery, also with a key in circuit, is joined to the box at A and B. The resistance R, which is to be measured, is placed between A and D.

When a measurement is to be made the sensitiveness of the galvanometer is reduced either by means of a shunt or by lowering the control magnet and 10 or 100 ohms taken from each of the ratio arms. Thus the value of P/Q is unity.

Some resistance, say 10 ohms, is then taken from the arm S, the battery circuit is closed, the galvanometer key depressed and the deflexion of the spot of light noted; let us suppose that it is to the right.

The value of S is then altered, let us suppose it is increased to 20 ohms, and the observation repeated; if the deflexion is still to the right but greater than before it is clear that the change in S has been in the wrong direction, it needs to be decreased; if the deflexion is less than before though still to the right it is clear that S is still too small. We will suppose the latter to be the case; increase S again and proceed thus until a deflexion to the left is obtained. After some few trials we shall be able to find two resistances which differ by one ohm, the smaller of which gives a deflexion to the right while the larger gives a deflexion to the left.

Let these resistances be 51 and 52 ohms. Then the required value of S lies between these two, and since

$$R = S \times \frac{P}{Q},$$

and $P/Q = 1$,

the value of R lies between 51 and 52 ohms.

Now remove the shunt or raise the control magnet to make the galvanometer more sensitive and make P 10 ohms, Q 100 ohms, then $P/Q = 1/10$ and the value of S required to give a balance will be 10 times what it was previously. It will therefore lie between 510 and 520. Proceeding as before we can find two resistances, 512 and 513 say, between which the required value of S must lie and since R is $S/10$ the value of R is between 51·2 and 51·3 ohms.

Proceeding thus, making P 10 and Q 1000, so that

$$P/Q = 1/100,$$

we can find two resistances, say 5125 and 5126 ohms, between which S must lie. Thus finally the value of R is between 51·25 and 51·26 ohms.

171. Resistance of a galvanometer. If a second galvanometer is available the resistance of a galvanometer coil can be measured like that of any other wire; it is sometimes convenient to use a galvanometer in the measurement of its own resistance.

EXPERIMENT 43. *To measure by Thomson's method the resistance of a galvanometer.*

Consider a Wheatstone's bridge arrangement in which the conjugate condition is satisfied so that $R/S = P/Q$, then the battery in AB produces no current in CD: it follows therefore that so far as the current in the other branches of the circuit is concerned it is immaterial whether CD is open or closed; if there be a key in CD, nothing in the rest of the circuit is altered by opening or closing this key; while conversely we may infer that if the currents in the rest of the circuit are not altered by opening or closing the circuit CD the conjugate condition is satisfied and

$$R = S \cdot \frac{P}{Q}.$$

To apply this, Fig. 168, the galvanometer whose resistance is to be measured is placed as the resistance R in the arm AD

and a key is placed in CD. The ratio arms P and Q are made
equal and a resistance taken out
in the arm S. On closing the
battery circuit the galvanometer
is deflected by the current in
AD.

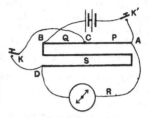

Fig. 168.

It is usually possible either
by inserting resistance in the
battery circuit or by the use of
the control magnet or of some
more powerful permanent magnet to bring the spot on to the
scale again. Let this be done and note its position. Now
make contact with the key K in CD; the spot will in general
be deflected, shewing that the current in AD is affected by
closing the circuit CD, thus the conjugate condition has
not been obtained.

By adjusting S however this deflexion can be made small,
and proceeding as in Experiment 42 we can find two resistances differing by 1 ohm for one of which there is a deflexion
to the left, for the other to the right. Thus R lies between
these two.

Now make P/Q equal to $1/10$ and find two other resistances
between which S lies as in Experiment 42. In this way a value
can be found for S which makes AB and CD conjugate and
thus R is equal to $S \cdot P/Q$.

It should be noted that the condition to be satisfied in this case is not
that there should be no current in the galvanometer but that there should
be no change in that current on depressing the key in the branch CD.

Moreover it must be remembered that there is a current in the
galvanometer which is changed in amount by each alteration of the
resistances P, Q or S; thus any alteration in the resistances will cause
a motion of the spot and it may need a corresponding change in the
control magnet to bring it back on to the scale.

It is desirable in this experiment that the battery used should be a
fairly constant one, otherwise changes in its E.M.F. when the key is depressed
may produce a motion of the spot, although the conjugate condition is
satisfied.

172. Resistance of a battery.

EXPERIMENT 44. *To find by Mance's method the resistance
of a battery.*

If we remember that the resultant effect of a number of electromotive forces is the sum of the effects due to the separate electromotive forces it is clear that we may place a battery in the arm AD of a Wheatstone's bridge without altering the conjugate condition that the current in CD should be independent of the E.M.F. in AB.

Suppose now that as shewn in Fig. 169 the battery whose resistance is required is placed in AD and a key in AB, the galvanometer being in CD.

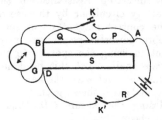

Fig. 169.

The battery will send a current through CD and the galvanometer will be deflected, if we can arrange the resistances so that this deflexion may be independent of the E.M.F. in AB; we know that AB and CD are conjugate, and hence that

$$\text{Battery resistance} = R = S \cdot \frac{P}{Q}.$$

Now when the conjugate condition is satisfied the current in CD is the same whatever be the E.M.F. in AD; take the case when this E.M.F. is zero, C and D are connected together through the key by a wire resistance; this will produce no effect on the galvanometer. Take again the case when the E.M.F. between A and B is exactly the amount required to balance that produced between these points by the current from the battery in AD, so that there is no current in AB; still the effect on the galvanometer is zero, this last condition must be the same as holds when the connexion between A and B is broken or the resistance between these points made infinite.

Thus if the resistances P, Q and S are so arranged that the galvanometer deflexion is the same whether the key in AB be depressed or not, we know the conjugate condition is satisfied.

To carry out this in practice the ratio arms are made equal

and a resistance taken out in the arm S; the galvanometer is deflected and the spot is brought on to the scale by shunting the galvanometer[1] if required, and by the use of the control magnet. The resistance S is then adjusted until the galvanometer deflexion is not altered by making or breaking the connexion in the branch AB by means of the key. As in the previous experiments two values are found for S, one of which gives a deflexion to the right, the other to the left, and the true value lies between these. The ratio P/Q is then altered to $1/10$ and the first decimal place in the value of R found as in Experiment 42.

As in Thomson's method an alteration of P, Q or S alters the permanent current through the galvanometer and the spot may need readjustment after each change in these resistances.

173. Specific Resistance. The relation between the resistance of a wire and its specific resistance or resistivity has been found in Section 140.

If l cm. be the length, a sq. cm. the area of the cross section, ρ the specific resistance and R the resistance of a wire. Then we have seen that

$$R = \frac{\rho l}{a},$$

and hence

$$\rho = \frac{Ra}{l}.$$

EXPERIMENT 45. *To measure by Wheatstone's bridge the specific resistance of the material of a wire.*

To find the specific resistance of a wire we have to measure its resistance R ohms, its length l cm., and the area of its cross section a sq. cm.

The resistance of the wire should if possible be several ohms, so as to render its accurate measurement possible. If sufficient wire is available, cut off a length having a resistance of 3 or 4 ohms at least. Measure the resistance carefully. Measure with a metre scale or steel tape the length of the wire; for this purpose it will be necessary

---

[1] Since we are measuring the battery resistance it is clear that we must not shunt the battery.

to lay out the wire in a straight line, but care must be taken not to stretch it.

Care must be also taken to measure exactly to the points in which the wire is clamped by the screws of the bridge; in some cases the wire may be soldered to thick pieces of copper of negligible resistance; these are connected to the bridge.

Measure by means of the wire gauge or in some other way the area of the cross section; this is most accurately done by finding the weight of water displaced by a known length of the wire; the weight of water displaced gives the volume of the wire[1], and by dividing the volume by the length, the area of the cross section is found. Then knowing the resistance, the length, and the cross section, the specific resistance is found.

Example. A piece of wire 1 metre in length is weighed in air and in water and the loss of weight is found to be ·825 gramme. Thus omitting corrections for the temperature of the water, the volume of the wire is ·825 c.cm., and the area of its cross section is ·00825 sq. cm. A length of 25 metres is measured off and is found to have a resistance of 6·05 ohms. Hence the specific resistance which is equal to Ra/l is ·00825 × 6·05/2500, and this comes to ·00001996 ohms, or 19·96 microhms for 1 c.cm.

[1] Glazebrook's *Hydrostatics* § 50, Exp. 14.

CHAPTER XVII.

MEASUREMENT OF QUANTITY OF ELECTRICITY, CONDENSERS.

174. The Ballistic Galvanometer. The quantity of electricity carried by a current is measured by the product of the current and the time of flow. In some cases a finite quantity is carried round a circuit in a very brief time, the current, measured by the rate of flow, is enormously great, but the time of flow being very small the quantity carried is finite; in such a case we can use a galvanometer to measure the quantity.

When a steady current is passing through the coils of a galvanometer, force is exerted on the magnet and the steady deflexion of the magnet becomes a means of measuring the current. The magnet is deflected until the force due to the current balances the force arising from the control magnet or, to put it rather differently, until the rate at which the current tends to produce momentum balances the rate at which the control magnet tends to produce it. The rate at which the current produces momentum is proportional to the current, and the impulse or total quantity of momentum produced is proportional to the product of the current and the time during which it has been flowing. But the product of the current and the time measures the quantity of electricity which has passed. Thus the impulse of the magnet is proportional to the quantity of electricity which has passed round the galvanometer coils. Now in some cases a finite quantity of electricity may traverse the galvanometer in a very short time; in such a case the current would be very large, but its time of flow very small; a definite quantity of momentum, proportional to the total quantity of electricity, is given to the magnet practically instantaneously, and this momentum will measure the quantity of electricity which has traversed the coils.

Hence in such a case the needle starts from rest with a momentum and therefore with an angular velocity which is a measure of the quantity of electricity which has traversed the coils. Now the motion of the needle resembles that of a pendulum under gravity; it is simple harmonic, and in such a case the velocity of the pendulum, as it passes its lowest point, is, when the swing is small, proportional to the amplitude of the swing[1].

Thus if a finite quantity of electricity traverse the coils of a galvanometer in a very short interval of time, the needle receives an impulse which is over before it has moved appreciably, and in consequence it swings out from its position of equilibrium, returning back through that position and finally after some few swings settles down to rest. The amplitude of the first swing is a measure of the quantity of electricity. It can be shewn[2] that if β be the angular magnitude of the first swing, T the time of vibration of the needle, and k the reduction factor of the galvanometer, then the approximate relation between Q, the quantity of electricity, and β is

$$Q = k \frac{T}{2\pi} 2 \sin \tfrac{1}{2}\beta,$$

$$= k \frac{T}{2\pi} \beta,$$

since β is very small.

Thus if we can arrange to discharge a quantity of electricity through the coils of a galvanometer in an interval of time so brief that we may assume the magnet has not moved appreciably from its equilibrium position before the flow ceases, we can measure the quantity by observing the first

[1] Glazebrook's *Dynamics* § 146.
[2] Glazebrook and Shaw's *Practical Physics*, Chapter XXI.

swing of the needle. If again a second quantity is discharged through the galvanometer in the same way the ratio of the two swings will give us the ratio of the two quantities.

Now when a condenser is charged there are equal quantities of positive and negative electricity respectively on its two plates. If these plates be connected through a galvanometer the condenser is discharged through the galvanometer, and this discharge takes place in a very brief interval. By observing the swing of the galvanometer we can measure the quantity passing in the discharge.

The galvanometer which is used for such experiments should have a long time of swing, for it is assumed that the time of discharge is short compared with the period ; it should also be arranged so as not to damp quickly, for the formula supposes that the magnet when once disturbed will continue to vibrate like a pendulum for some time ; if however the experiments are merely relative, this last requirement loses its importance, for the damping affects all the swings in the same proportion.

175. Condensers. We proceed to describe some experiments on condensers in which this ballistic method of measuring a quantity of electricity is made use of.

In these and for many other purposes a Morse key, Fig. 170, is useful.

The handle of the key is a lever carrying two contact pieces which make contact with two studs on the base of the key. These studs are connected to two binding screws K_1, K_2. The fulcrum on which the lever works is

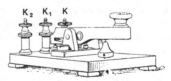

Fig. 170.

attached to a third binding screw K. A spring keeps the handle raised so that in the normal position there is connexion between K and K_1. On depressing the handle this connexion is broken and contact is established between K and K_2.

EXPERIMENT 46. *To compare by means of a galvanometer the capacities of two condensers.*

A battery B of constant E.M.F.—*e.g.* one or two Daniell's cells, or preferably a storage cell—a condenser A and a galvanometer G are connected through a Morse key as shewn in Fig. 171.

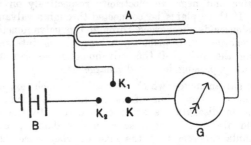

Fig. 171.

With the key in the normal position K and K_1 are in contact and the two plates of the condenser are connected through the galvanometer; when the handle is depressed K_1 is insulated, while K_2 is put into connexion with K, the condenser is charged through the galvanometer and the first throw of the needle is proportional to the charge. Observe the throw and wait until the needle has again come to rest, then release the key. The condenser is discharged through the galvanometer and the throw, in the opposite direction to that previously noted, measures the discharge. If there be no leakage and no electric absorption the throws will be equal.

Take a series of readings on the galvanometer scale of the throws due to charge and discharge, and let the mean of these be b_1. If the throw is small, b_1 is approximately proportional to the angular deflexion β_1, and therefore to the quantity of electricity which has passed. Let this quantity be Q_1.

Replace the first condenser by the second and repeat the observation. Let Q_2 be the charge of the condenser, using the same battery, and let b_2 be the mean of the throws as measured on the scale.

Then Q_2 is proportional to b_2.

Hence $\qquad\qquad Q_1 : Q_2 = b_1 : b_2$.

But if E be the E.M.F. of the battery, C_1, C_2 the capacities of the two condensers, we have

$$Q_1 = EC_1, \quad Q_2 = EC_2.$$

Thus $\qquad\qquad C_1 : C_2 = b_1 : b_2$

and the two capacities can be compared.

EXPERIMENT 47. *To compare by the aid of a condenser the electromotive forces of two batteries.*

The measurements are the same as in the last experiment but instead of changing the condenser we use the same condenser with two different batteries.

Let E_1, E_2 be the electromotive forces of the batteries, C the capacity of the condenser, so that now Q_1 is the charge if the condenser is charged to a potential difference E_1, Q_2 its charge when the potential difference is E_2.

As before $\qquad\qquad Q_1 : Q_2 = b_1 : b_2$.

But $\qquad\qquad Q_1 = CE_1, \quad Q_2 = CE_2$.

Hence $\qquad\qquad \dfrac{E_1}{E_2} = \dfrac{b_1}{b_2},$

and thus the electromotive forces are compared.

EXPERIMENT 48. *To measure by means of a ballistic galvanometer the capacity of a condenser.*

The connexions are made as in Fig. 172, R is a high resistance and LMN a two-way switch; K, K_1, K_2 being a Morse key. In the normal position K and K_1 are in connexion and if L and M are connected the condenser is charged, the charge however does not traverse the galvanometer. On depressing the key the condenser is discharged through the galvanometer and if Q be the amount of the charge, β the first throw of the needle, T the time of swing and k the reduction factor, then, as we have seen in Section 174,

$$Q = k\,\frac{T}{2\pi} \times \beta = EC.$$

Now release the Morse key and shift the switch contact from M to N. A current from the battery can flow through the resistance R and the galvanometer of resistance G, the

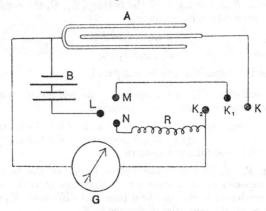

Fig. 172.

amount of this current is $E/(B + G + R)$ and B the battery resistance is usually negligible compared with R. Thus if θ be the steady deflexion of the needle we have

$$\frac{E}{G+R} = k \tan \theta = k\theta,$$

since θ is small.

Thus $E = k\,(G + R)\,\theta.$

Hence $C = \dfrac{T}{2\pi\,(G + R)}\,\dfrac{\beta}{\theta},$

and since the quantities on the right are known or can be observed the capacity C is determined.

In practice it will be found that unless R is very large the steady deflexion will be too great to measure with a galvanometer sufficiently large to give a measurable throw when the condenser is discharged; it may often be necessary to shunt the galvanometer for the second part of the experiment and in this case the formula needs a suitable modification[1].

[1] For further details as to these experiments reference should be made to Glazebrook and Shaw's *Practical Physics*.

176. Unit of capacity. The capacity of a condenser is measured in terms of a unit of capacity, and since capacity is the ratio of the charge to the difference of potential, a condenser in which unit charge produces unit potential difference will have unit capacity. Experiment shews that, if we take as the unit charge and the unit potential difference the electromagnetic units, then the condenser of unit capacity would be an enormously large piece of apparatus. It is more convenient to adopt as the unit the capacity of a condenser which when charged with 1 coulomb has a potential difference between its plates of 1 volt. This unit is known as a Farad (from Faraday). Even this unit however is far too large for ordinary use.

DEFINITION. *The capacity of a condenser in which a charge of* 1 *coulomb produces a potential difference of* 1 *volt is taken as the practical* Unit of Capacity *and is called* 1 Farad.

Since 1 coulomb is 1/10th of the C.G.S. unit of electricity and 1 volt is 10^8 C.G.S. units of potential, a condenser whose capacity is 1 farad contains $1/10^9$ or 10^{-9} C.G.S. units of capacity.

A farad however is found to be too big a unit for most purposes and as a unit for ordinary use a **Microfarad** or one-millionth of a farad is taken.

A microfarad $= 10^{-6}$ farad $= 10^{-15}$ C.G.S. units of capacity.

If a condenser having a capacity of 1 microfarad is charged with 1 coulomb the potential difference between its plates would be one million volts.

CHAPTER XVIII.

THERMAL ACTION OF A CURRENT.

177. Heating of a conductor by a current.
We have seen in § 113 that a conductor carrying a current
becomes heated. This heat is the equivalent of the energy
lost by the current in passing along the conductor in the
direction of the fall of the potential.

We have also seen, § 37, that when a quantity of electricity
Q passes from a point at potential V_1 to a point at lower
potential V_2 electrical energy equal to $QV_1 - QV_2$ disappears.
In a case in which the electricity in its flow does no external
work this energy, which may be written $Q(V_1 - V_2)$, reappears
as heat in the conductor.

If we write E for the fall of potential $V_1 - V_2$ between the
points where the current enters and leaves the conductor,
then the loss of energy will be EQ, and if the transfer be
due to a current of strength i flowing for t seconds $Q = it$.
Thus if W be the work done, measured by the loss of electrical
energy, we have

$$W = EQ = E \cdot i \cdot t.$$

Now let H be the heat generated in the conductor and J
Joule's equivalent[1]. Then since each unit of heat is the

---

[1] Joule's equivalent is the amount of energy measured in ergs which,
if entirely transformed into heat, would raise 1 gramme of water 1° C.
Glazebrook's *Heat*.

equivalent of J units of work, and since an amount of work W is converted into heat H we have $W = JH$.

Thus $\qquad\qquad JH = E \cdot i \cdot t$;

or if we write H' for the rate at which heat is produced, *i.e.* the amount of heat generated per second, then $H' = H/t$ and hence

$$JH' = Ei.$$

178. Joule's Law. The above result is known as Joule's law.

By combining it with Ohm's law we can put it in various forms. For from Ohm's law $E = Ri$ or as we may write it $i = E/R$ where R is the resistance of the conductor between whose ends a difference of potential E is maintained.

Thus $\qquad\qquad JH' = Ei = Ri^2 = \dfrac{E^2}{R}.$

Hence if a given current flow through a wire of given resistance, the heat produced is proportional to the product of the resistance and the square of the current, while if a given E.M.F. is applied to the ends of the wire the heat produced is proportional to the square of the E.M.F. divided by the resistance.

These laws can be verified by various experiments.

EXPERIMENT 49. *To verify Joule's law.*

A piece of thin silk-covered wire of German-silver or some other resistance material having a resistance of 4 or 5 ohms is wound into a spiral; the ends of the wire are connected to two thick leads of copper; the wire can be immersed in water in a small copper calorimeter (Fig. 173), the leads passing through the cork which closes the calorimeter; a stirrer and a thermometer also pass through the cork. The coil forms part of a circuit which contains a battery of some 5 or 6 volts E.M.F., an ammeter and a key.

The calorimeter contains a known mass M grammes of water; a resistance box may conveniently be included in the circuit to vary the current or if preferred this can be done by altering the number of cells. The temperature of the water is taken, then the circuit is closed and the current allowed to pass for 10 minutes. The strength of the current is noted on the ammeter and the rise of temperature observed, the water

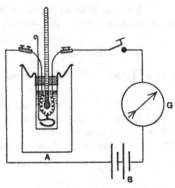

Fig. 173.

being well stirred. At the end of the ten minutes the current is stopped, the temperature continues to rise for a short time and the highest point reached is observed. Thus the total rise is found. Let this be $T°$ C. Then neglecting the loss by radiation and the heat used in raising the temperature of the calorimeter the heat given out by the wire has raised M grammes of water $T°$, its amount therefore is MT and this has been produced by a current of strength i amperes flowing for 10 minutes.

Now cool[1] the water down to approximately the same temperature as it had at the beginning, vary the current, and repeat the experiment. Allowing the new current i' to flow for 10 minutes, a rise of temperature $T'°$ will be observed; and it will be found that we have as approximately true the relation

$$\frac{i^2}{i'^2} = \frac{T}{T'}.$$

[1] By this means the loss from radiation is made to affect the two experiments more nearly to the same extent than would be the case if the second experiment were started at the temperature at which the first finishes.

The rise of temperature, since the mass of water and other conditions are constant, is proportional to the heat caused by the current, and is found to vary as the square of the current.

EXPERIMENT 50.   *To apply Joule's law to the measurement of electromotive force.*

We have from Joule's law the relation

$$JH = Eit,$$

where H is the heat produced in t seconds.

Hence
$$E = \frac{JH}{i \cdot t}.$$

The previous experiment has enabled us to measure the various quantities on the right-hand side with the exception of J. We have seen (Glazebrook, *Heat*, § 176) how this constant can be measured, and its value has been shewn to be 42×10^6 ergs. The value[1] of H is MT where M is the mass and T the rise of temperature of the water. The current i should be measured in C.G.S. units, since the other quantities are in C.G.S. units, hence if the ammeter reads directly in amperes and if A be the reading observed $i = A/10$.

Hence
$$E = \frac{4 \cdot 2 \times MT}{At} 10^8;$$

t is the time of flow of the current which we have taken to be 10 minutes.

The value of E is given in C.G.S. units of electromotive force. Since 1 volt $= 10^8$ C.G.S. units, in order to find the value in volts we must divide by 10^8 and we thus get

$$E = \frac{4 \cdot 2 \times MT}{At} \text{ volts.}$$

Example. *A current of ·75 ampere flows through the wire for 10 minutes. The mass of water in the calorimeter is 85 grammes and the rise of temperature 5°·4 C. Find the potential difference between the ends of the wire.*

---

[1] This value requires correcting (1) for the loss by radiation, (2) for the heat given to the calorimeter and stirrer: for the method of applying these corrections, see Glazebrook's *Heat* and Glazebrook and Shaw's *Practical Physics*.

We have $M = 85$, $T = 5°·4$, $A = ·75$, $t = 600$ seconds.

Hence $$E = \frac{4·2 \times 85 \times 5·4}{450} = 4·28 \text{ volts.}$$

Again from the above we can calculate the resistance of the wire in which the current is flowing, for the current of ·75 ampere is produced by an E.M.F. of 4·28 volts.

Hence the resistance is 4·28/·75 or 5·71 ohms.

179. Joule's law applied to a battery.

It should be noticed that in the above sections E stands for the difference of potential between the ends of the wire and R the resistance of the wire.

We can however apply the same reasoning to the complete circuit including the battery. If E be the electromotive force of the battery a quantity of electrical energy EQ disappears when Q units of electricity pass completely round the circuit, and if no external work is done this reappears as heat in the various conductors of which the circuit is composed. If we write $Q = it$ and $E = (B + R)i$, B being the battery resistance, R that of the rest of the circuit, then

$$H = \frac{(B + R)i^2 t}{J},$$

and of this heat an amount $Bi^2 t/J$ appears in the battery, $Ri^2 t/J$ in the external part of the circuit.

Example. *A battery whose resistance is half an ohm is producing a current of 15 amperes in a circuit whose resistance is 10 ohms, find the heat generated per second in the battery and in the circuit. Find also the E.M.F. of the battery.*

We have $E = Ri = 15 \times 10·5 = 157·5$ volts.

In calculating the heat we must remember that in this formula both i and R are to be measured in C.G.S. units.

Hence $\qquad i = 15$ amperes $= 1·5$ C.G.S. units,

$\qquad\qquad R = 10$ ohms $= 10 \times 10^9$ C.G.S. units,

$\qquad\qquad B = ·5$ ohm $= ·5 \times 10^9$ C.G.S. units,

$\qquad\qquad J = 4·2 \times 10^7$ ergs.

Hence for the battery

$$\text{Heat} = \frac{1·5 \times 1·5 \times ·5 \times 10^9}{4·2 \times 10^7} = 26·8 \text{ units,}$$

and for the wire

$$\text{Heat} = \frac{1·5 \times 1·5 \times 10 \times 10^9}{4·2 \times 10^7} = 535·6 \text{ units.}$$

The total heat is given by the formula

$$\text{Heat} = Ei/J,$$

and since we have seen that $E = 157 \cdot 5$ volts, $i = 15$ amperes, we obtain

$$\text{Heat} = 562 \cdot 5 \text{ units}$$

and this is very nearly[1] the sum of $26 \cdot 8$ and $535 \cdot 6$.

180. Electrical Energy. Whenever a quantity Q of electricity is transferred round a circuit in which there is an E.M.F. E, the work done as measured by the electrical energy is EQ; if E and Q are both measured in C.G.S. units this electrical energy is measured in ergs; it is more convenient however to measure the E.M.F. in volts, the current in amperes and the quantity transferred in coulombs, to adopt, that is, the practical rather than the C.G.S. system of units.

The unit of energy on this system will be the work done when 1 coulomb is conveyed round a circuit under an E.M.F. of 1 volt. This unit is called a **Joule**.

DEFINITION. *A* **Joule** *is the amount of electrical energy expended by the transference of* 1 *coulomb round a circuit in which the electromotive force is* 1 *volt.*

Since 1 coulomb $= \frac{1}{10}$ C.G.S. unit and 1 volt $= 10^8$ C.G.S. units, we see that

$$1 \text{ joule} = 10^7 \text{ C.G.S. units of work} = 10^7 \text{ ergs.}$$

Moreover since Joule's equivalent has been shewn to be $4 \cdot 2 \times 10^7$ ergs, we see that

$$\text{Joule's equivalent} = 4 \cdot 2 \text{ joules.}$$

181. Electrical Power. Power has been defined[2] as the rate of doing work, it is measured therefore by the number of units of work done per second. In C.G.S. units it is the number of ergs done per second.

It is convenient to have a practical unit of power related to the joule. This is called the **Watt** and is the power expended when work is being done at the rate of 1 joule per second.

---

[1] The difference is due to not carrying the figures in the arithmetic to a sufficient number of decimal places.

[2] Glazebrook, *Dynamics*, § 110.

One joule is done if a coulomb is transferred round the circuit under an E.M.F. of 1 volt. Now if a coulomb is transferred per second the current is 1 ampere, hence if a current of 1 ampere is flowing under an E.M.F. of 1 volt the power expended is 1 watt.

DEFINITION. *A power of* 1 **Watt** *is expended in producing a current of* 1 *ampere under an electromotive force of* 1 *volt.*

Moreover since 1 ampere = 1/10 C.G.S. unit of current and 1 volt = 10^8 C.G.S. units of E.M.F.

1 watt = 10^7 C.G.S. units of power = 10^7 ergs per second.

Thus in any given circuit the power in watts is found by multiplying the volts by the amperes, while the work done in a given time is found in joules by multiplying the power in watts by the time in seconds.

A unit of power often employed in practice is the kilowatt, or 1000 watts. This is known as the Board of Trade unit.

The relation between the erg and the foot-pound is known, and it can be shewn that[1]

$$1 \text{ erg} = \cdot737 \times 10^{-7} \text{ foot-pound.}$$

Thus it follows that

$$1 \text{ joule} = \cdot737 \text{ foot-pound,}$$

and $\qquad$ $1 \text{ foot-pound} = 1\cdot356 \text{ joules.}$

Again when work is done at the rate of 550 foot-pounds per second 1 horse-power is exerted.

Hence

$$1 \text{ horse-power} = 550 \times 1\cdot356 \text{ joules per second}$$

$$= 746 \text{ watts.}$$

$$1 \text{ watt} = 1/746 = \cdot001340 \text{ horse-power.}$$

$$1 \text{ kilowatt} = 1\cdot34 \text{ horse-power.}$$

EXPERIMENT 51. *To determine Joule's equivalent by electrical measurements.*

The experiments described in Section 177 can be utilized to find Joule's equivalent. For we have the equation $JH = Eit$.

---

[1] Glazebrook, *Dynamics*, § 110.

Now E and i can be found by electrical measurements, and H is determined as in Experiment 50, by the calorimeter, and then from the equation $J = Eit/H$ we find J.

182. Thermo-electricity. If a circuit be composed of two different metals, and if one of the two junctions of the metals be at a different temperature to the other, a current is produced round the circuit.

This current which may be shewn in various ways is said to be due to **Thermo-electric action.**

EXPERIMENT 52.   *To shew the thermo-electric production of a current.*

(*a*) A strip of copper and one of German-silver are brazed together at their ends and bent so as to form a narrow rectangle, as shewn in Fig. 174. This is placed with its long edges horizontal and a compass-needle is mounted so that its centre may coincide with that of the rectangle.

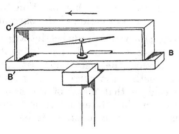

Fig. 174.

The plane of the rectangle lies in the meridian, parallel to the axis of the compass-needle. On heating one of the junctions with a Bunsen-burner it will be noticed that the compass-needle is deflected ; a current is traversing the circuit. Note the direction of the deflexion and infer thence the direction of the current.

(*b*) Two pieces of copper-wire are fastened to the ends of an iron rod. This may be done by drilling a hole through the rod, passing the copper-wire through the hole and winding it tightly for a few turns round the rod. The other ends of the wires are connected to an ammeter or low resistance galvanometer; one of the junctions is heated by a Bunsen-burner and the galvanometer needle is deflected. Notice the direction of the deflexion and so find that of the current.

183. Observations on Thermo-electricity. It will be found that if in the last experiment one junction be kept at the ordinary temperature while the other is gradually heated, the current will at first pass from copper to iron through the hot junction, and this current will rise as the temperature of the hot junction is raised until that temperature reaches a limiting value T, known as the neutral temperature for these two metals. In the case of iron and copper this neutral temperature is about $284°$ C. When the temperature of the hot junction is raised above the neutral temperature the E.M.F. in the circuit and in consequence the current decrease, and this continues until the hot junction is as much above the neutral point as the cold junction is below, when the current is again zero; if the hot junction be still further raised in temperature the direction of the current is reversed.

The thermo-electric force produced by a given difference of temperature varies very much for different metals, and it is possible to arrange the metals in a list so that if two of them are joined together the current will pass from the hot junction to the cold in that metal of the two which stands last on the list. In this statement it is assumed that the mean temperature of the two junctions is below their neutral temperature.

Thus at ordinary temperatures bismuth, German-silver, lead, platinum, copper, zinc, iron, antimony form such a list. In thermo-electric experiments it is usual for a reason which will appear later to take lead as the standard metal and to refer other metals to it. As the temperature of the hot junction rises the electromotive force changes, and the rate of increase of the E.M.F. for each rise of temperature of $1°$ C. at any given temperature is called the thermo-electric power at that temperature. In the case of metals which stand before lead on the list the thermo-electric power is positive; the thermo-electric force is from the hot junction to the cold and is increased by a rise of temperature; in the case of metals below lead the thermo-electromotive power is negative; the thermo-electromotive force from the hot junction to the cold increases negatively with rise of temperature, that is, the electromotive force is as in the case of the copper-iron junction from the cold junction to the hot.

By measuring the E.M.F. in a given circuit in which one junction is kept at some convenient temperature, say 0° C., while the other is heated, and plotting the results we can obtain a curve in which the ordinates give the thermo-electromotive force and the abscissæ the temperature, and from these curves the thermo-electromotive power which is the rate at which the E.M.F. increases with the temperature can be calculated. Thus a second series of curves of thermo-electromotive power can be plotted.

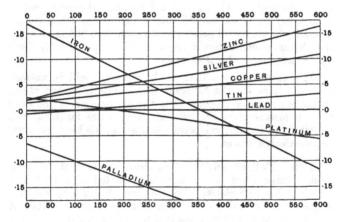

Fig. 175.

For a large number of metals and over a considerable range of temperature these curves of thermo-electric power are found to be straight lines. The same experiments shew that the curves of thermo-electric force are parabolas.

Fig. 175 gives a thermo-electric diagram or series of

curves of thermo-electric power, the abscissæ represent
temperatures, the ordinates give the thermo-electric power
measured in microvolts per degree of temperature.

The thermo-electric power at any temperature will measure
very approximately the E.M.F. round a circuit of the given
metal and lead when one junction is $\frac{1}{2}$ a degree above, the
other $\frac{1}{2}$ a degree below, the given temperature. When the
thermo-electric power is positive it means that the current
passes from the hot junction to the cold in the lead. Thus
bismuth is thermo-electrically positive with regard to lead
since the current passes at the hot junction from the bismuth
to the lead. Antimony again is thermo-electrically negative,
for at the hot junction the current is from lead to antimony.

The electromotive forces caused by heating a single junction
are very small; thus if one junction of a copper-iron circuit
be at 1° C., the other being at 0° C., the E.M.F. will be about
15 microvolts[1]; in the case of a bismuth-antimony circuit it
is greater, being about 110 microvolts; some other metals
and alloys have a still greater value than this, but in all cases
the thermo-electromotive force is small compared with those
which arise from chemical action.

If a thermo-electric diagram be drawn it will be noticed
that the lines for the different metals intersect each other,
the temperature corresponding to the intersection of the lines
of two metals gives the neutral temperature for those two
metals. At the neutral point the two metals have the same
thermo-electric power.

184. Peltier effect. The discovery that there is an
electromotive force in an unequally heated circuit of two
metals is due to Seebeck. The converse of this fact, viz. that
if a current be made to flow across the junction it heats it if
it flows in one direction, cools it if it flows in the other, is due
to Peltier. Seebeck shewed that in a copper-iron circuit a
current will pass from copper to iron across the heated
junction. Peltier proved that if a current be made to pass
from copper to iron across a junction that junction is cooled,
if however the current flows from iron to copper the junction

---

[1] 1 microvolt = one millionth of a volt = 10^{-6} volt.

is heated. This heating effect must be distinguished from the Joule effect; the resistance of the circuit causes the evolution of heat whichever way the current flows; the Peltier effect is reversible with the current; in one direction heat is evolved with the current, in the other it is absorbed and the junction is cooled.

The Peltier effect is more marked in two metals such as bismuth and antimony which are at some distance apart in the thermo-electric series than between two such as copper and iron.

When a bismuth-antimony junction is heated the current flows from bismuth to antimony; if a current be passed across the junction in this direction the junction is cooled. This can be proved by enclosing one of the two junctions in the bulb of an air thermometer as in Fig. 176. The two metals pass through tubulures at either side of the bulb and are sealed into it with some air-tight cement. On passing a current from antimony to bismuth the junction is heated and the column of liquid in the thermometer tube falls; on reversing the current the junction is cooled and the column rises. It must be noted however that the metal is being heated in consequence of its resistance; this heats the air in the bulb, and unless this heating be less than the cooling due to the Peltier effect the column will not rise in the latter case, but will fall more slowly than in the former. This difficulty may be overcome by the use of a differential air thermometer

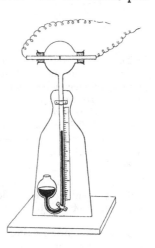

Fig. 176.

as shewn in Fig. 177, the one junction is in one bulb, the other in the second bulb. The Joule heating effect is the same for the two and the motion of the liquid will depend on the difference of the heating effect in one bulb and the cooling effect in the other.

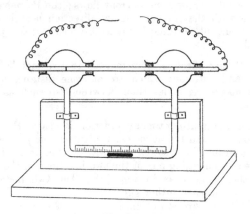

Fig. 177.

The heat produced at the junction is found to be proportional to the current passing and to the time, *i.e.* to the quantity of electricity which has crossed the section. Thus if i be the current in C.G.S. units and t the time in seconds the energy absorbed or evolved at the junction may be written $P.i.t$, and P is a coefficient which is called the coefficient of the Peltier effect and is measured by the energy evolved by the passage of 1 C.G.S. unit of electricity.

Moreover it can be shewn by aid of the thermo-electric diagram that the thermo-electric power at any junction is found by dividing the coefficient of the Peltier effect by the temperature of the junction measured from absolute zero.

185. Thomson effect. It was shewn by Lord Kelvin that the passage of a current along an unequally heated conductor of any material except lead causes the absorption or evolution of heat at each point of the conductor according to the direction of the current. In the case of copper heat is absorbed and the wire cooled if the current flows from the cold part of the wire to the hot part; in the case of iron the reverse is true; this is known as the Thomson effect. If the direction of the current be reversed the absorption of heat becomes an evolution and conversely.

The absence of the Thomson effect in lead is the reason why lead is chosen as the standard metal in thermo-electric measurements.

186. The Thermopile. The thermo-electromotive force of an antimony and bismuth couple is made use of in a thermopile for the measurement of small differences of temperature. A number of bars of these metals are arranged alternately, as in Fig. 178, A 1, 2 3, 4 5, being antimony bars and 1 2, 3 4, 5 B bismuth bars; the bars are soldered together at 1, 2, 3, 4, 5, the ends A and B being connected to a low resistance galvanometer. If the junctions 1, 3, 5 be heated while 2 and 4 remain cool a current will flow through

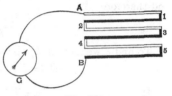

Fig. 178.

the galvanometer from A to B. If the junctions 2, 4 be heated, 1, 3, 5 remaining cool, the direction of the current will be reversed. The E.M.F. is proportional to the number of junctions, hence in the apparatus as usually made a large number of junctions are connected up in square order, as shewn in Fig. 179, the contiguous bars of metal being insulated from each other electrically by strips of mica.

Fig. 179.

The same principle is applied to the measurement of high temperatures; a couple made of wires of pure platinum and an alloy of platinum-rhodium or platinum-iridium is employed. By means of preliminary experiments, either by a comparison with a standard thermo-couple or by the aid of a suitable air thermometer, a curve giving the relation between E.M.F. and temperature is plotted for the couple. After this has been done a determination of the E.M.F. under any conditions will give the temperature of the junction.

187. Platinum Thermometer. Another electrical method of measuring temperature depends on the fact that the resistance of a wire rises as its temperature is increased. If then we know (1) the resistance of a piece of wire at some given temperature 0° C. (say), and (2) the rate at which the resistance increases with the temperature, then we can by measuring the resistance of the wire determine its temperature.

For if R_0 be the resistance at 0° C. and aR_0 the change in resistance for each degree centigrade; then assuming for the present that a is a constant the increase of resistance for t° C. is aR_0t. Hence if R be the resistance at t° we have

$$R = R_0 + R_0at,$$

and therefore

$$t = \frac{R - R_0}{aR_0}.$$

To find a we have if R_{100} be the resistance at 100° C.

$$R_{100} = R_0 + aR_0 \cdot 100.$$

Hence

$$a = \frac{R_{100} - R_0}{100R_0}.$$

Thus

$$t = 100\frac{R - R_0}{R_{100} - R_0}.$$

The wire usually employed for the purpose is platinum which is carefully annealed; now it has been shewn by Professor Callendar that the coefficient of increase of resistance of platinum per degree centigrade is not the same for all temperatures, so that if the symbol t in the above formulæ is taken to mean temperature in degrees centigrade they are not exactly true. We may however adopt a scale of temperature for which they are exact; this is known as the platinum scale and a rise of 1° on the platinum scale is a rise of temperature which produces a change in the resistance of the wire of one hundredth the amount occurring between the freezing point and the boiling point.

If we agree to reckon temperatures by this platinum scale then the equation

$$t_{pt} = 100\frac{R - R_0}{R_{100} - R_0}$$

is true.

Moreover Professor Callendar has shewn that there is a very simple connexion between temperature measured on the platinum scale and temperature measured on the centigrade scale, which is given by the equation

$$t - t_{pt} = \delta \cdot \frac{t_{pt}}{100} \left(\frac{t_{pt}}{100} - 1 \right),$$

where δ is a constant, depending on the wire, which for pure platinum differs very little from the value 1·5.

The wire employed as a thermometer is coiled on a mica frame and enclosed inside a glass or porcelain tube which can be immersed in the material whose temperature is to be measured in the same way as an ordinary thermometer; the resistance of the wire is then measured; this gives us R, and hence R_0 and R_{100} being known from preliminary observations we can find t_{pt}; then t is given by the formula. With the usual value of δ the difference between t and t_{pt} at 50° C. is 0°·37 while at 500° C. it is about 30°.

There is an arrangement whereby compensation is secured for the resistance of the connexions leading to the coil so that the resistance of the platinum spiral alone is measured.

Fig. 180.

Fig. 180 shews a platinum thermometer as usually constructed.

CHAPTER XIX.

THE VOLTAIC CELL. (Theory.)

188. Energy changes in a cell. We have already seen, § 115, that if two platinum plates are immersed in acidulated water and a potential difference maintained between them, then positive electricity flows from the plate at higher potential to that at lower, while oxygen is deposited at the first plate, hydrogen at the second, and moreover that for each gramme of hydrogen deposited on the latter, 8 grammes of oxygen are deposited on the former. Again if two metallic plates of different materials are placed in dilute sulphuric acid, let us say copper and zinc, and connected by a copper wire outside the acid, then positive electricity passes through the acid from the zinc; hydrogen collects on the copper and for each gramme of hydrogen deposited, 32·5 grammes of zinc are taken from the zinc plate which combine with 48 grammes of sulphion, SO_4, to form 80·5 grammes of zinc sulphate.

Now it is from this combination of the zinc and acid that the energy required to drive the current is produced and we can obtain a relation between the electromotive force of the battery and the energy of chemical combination thus.

Let us call H the energy set free when one gramme of zinc combines with oxygen; it is known as the heat[1] of

---

[1] We here suppose the heat to be measured as energy : if we take h as the heat of combination in calories and J as Joule's equivalent, we have $H = Jh$.

combination of zinc and oxygen, and can be determined from calorimetric observations. Then if a quantity Q of positive electricity pass from the zinc to the acid, and if γ be the electrochemical equivalent of zinc, a mass of zinc $Q\gamma$ grammes is removed and sets free an amount of energy $HQ\gamma$.

The result is the transference of a quantity Q round the circuit, and if E be the E.M.F. of the battery the work needed to do this is EQ units of work.

If we assume that this work is derived from the chemical combination, then we must have

$$EQ = HQ\gamma$$

or $E = H\gamma$.

We thus express the E.M.F. of the cell in terms of the heat of combination of the zinc and oxygen and the electro-chemical equivalent of the metal.

This simple theory requires, as Helmholtz shewed, some modification from the fact that the E.M.F. of a cell depends on its temperature, while in consequence of the passage of the current reversible thermal changes go on at various junctions, but in most cases in practice the correction introduced by this consideration is small.

189. Electromotive Force of a Daniell's Cell. In the above we have assumed that all the chemical inter-changes occur at the zinc plate; the whole energy of the cell comes in this case from this change. In reality in most cells changes occur elsewhere which give rise to the absorption or liberation of energy, and when all these changes are taken into consideration the calculated E.M.F. is found to agree closely with that observed.

Thus if we suppose that in a Daniell's cell energy is liberated by the removal of the zinc and absorbed by the deposition of the copper, and that no other chemical actions involving liberation or absorption of energy occur, we proceed to calculate the E.M.F. as follows.

It has been shewn that the combination of 32·5 grammes of zinc to form zinc sulphate liberates 54231 calories, while the deposition of the corresponding amount 31·6 grammes of copper from copper sulphate absorbs 27112 calories, thus the heat equivalent of the energy liberated on the passage of the quantity of electricity required to dissolve this quantity of zinc is the difference, or 27119 calories.

This quantity of electricity deposits 1 gramme of hydrogen, and since the c.g.s. unit of electricity deposits 0·0001038 grammes of hydrogen, the quantity of electricity required to deposit 1 gramme is 1/·0001038 or 9634 c.g.s. units.

Thus the quantity of electricity which has passed is 9634 c.g.s. units.

Now since the value of the mechanical equivalent of heat is $4·2 \times 10^7$ ergs, the energy liberated is

$$27119 \times 4·2 \times 10^7 \text{ ergs,}$$

and the E.M.F. produced being the energy liberated on the passage of unit quantity of electricity is

$$27119 \times 4·2 \times 10^7/9634 \text{ c.g.s. units.}$$

This it will be found reduces to $1·18 \times 10^8$ c.g.s. units, or since 1 volt is 10^8 units we have for the E.M.F. of the cell the value 1·18 volts. We have thus calculated the E.M.F. of the cell from the chemical changes.

The result is some 5 or 6 per cent. greater than the value found by direct experiment, but the latter depends on the concentration of the zinc sulphate and in the theoretical account various small corrections have been omitted.

190. Chemical and Electrical Transformations in Electrolysis.

Let us now endeavour to picture to ourselves what is going on when acidulated water is being decomposed in an electrolytic cell, or, when a voltaic cell is producing a current, and see if we can arrive at any satisfactory theory of the action of the cell.

A molecule of water consists of two atoms of hydrogen and one of oxygen. With the water are molecules of sulphuric acid H_2SO_4. Under the action of the electric forces these are decomposed; the hydrogen atoms carrying positive electricity move with the positive current to the negative plate. Sulphion, SO_4, carrying a negative charge is set free, this re-combines with two atoms of hydrogen from the water to form sulphuric acid H_2SO_4 setting free an atom of oxygen with its corresponding negative charge to appear at the positive plate.

Again since in the electrolysis of water the masses of oxygen and hydrogen deposited in a given time are in the ratio of 8 to 1, while the masses of the respective atoms are as 16 to 1, it follows that the number of hydrogen atoms deposited in that time is twice as great as that of oxygen atoms : but since the quantities of positive and negative electricity concerned are equal the charge of each oxygen atom, if we suppose the electricity carried by the atoms, is twice as great as that of an hydrogen atom ; this is found to be the case for any divalent constituent of a salt.

Hence if we call e the charge of an atom of hydrogen, the charge of the atom of oxygen is $-2e$, that of the group SO_4 is also $-2e$, while since in copper sulphate or in zinc sulphate the atom of copper or zinc replaces respectively two atoms of hydrogen the charge is in each case $+2e$. In the case of silver nitrate however, which is univalent, the charge on each atom of the silver is $+e$.

Thus we are to look upon sulphuric acid as a series of molecules H_2SO_4 in which each of the hydrogen atoms carries a charge $+e$, while the SO_4 group carries $-2e$; the whole charge therefore is zero, being made up of $+2e$ on the two hydrogen atoms and $-2e$ on the SO_4.

It may possibly be better to consider the group SO_4 as SO_3 with a charge $-e$ and O also with a charge $-e$.

When electric force acts on the liquid some of the molecules are apparently split up into H_2 with $+2e$ and SO_4 with $-2e$.

There are however a number of facts which point to the conclusion that the union between two or more atoms to form a molecule is not a permanent one, but that continuous interchanges of partners are always going on among the molecules. At any moment by far the greater number of the atoms are combined into molecules, but there are a certain number of free hydrogen atoms with charge e and half as many of the

SO_4 groups with charge $-2e$. These free atoms and groups are known as ions. When the electric force acts the free hydrogen ions are carried in the direction of the force; some of them combine, on the way, with the free sulphion groups which with their negative charges are moving in the opposite direction; some reach the negative electrode and give up to it their positive charge. The sulphion groups combine with the hydrogen of the water molecules, thus setting free oxygen ions with negative charges which travel to the positive plate; in this way we can explain the main phenomena of electrolysis.

191. Chemical and Electrical Transformations in a Voltaic Cell. Let us now consider how we can apply these ideas to the phenomena of the cell.

In accordance with them when a plate of pure zinc is placed in dilute acid the zinc begins to combine with the acid and form zinc sulphate. Each atom of zinc carries with it its charge $+ 2e$ and leaves the zinc plate negatively charged.

We here assume that zinc consists of an equal number of positively and negatively charged atoms; when the positive atom combines with the SO_4 the negative charge is set free on the zinc plate.

But corresponding to each negative ion SO_4 which has combined with the zinc two positive hydrogen ions have been left free in the liquid, these are attracted to the negatively charged zinc and form a coating round it, and a double sheet is produced about the zinc consisting of the negatively charged zinc ions overlaid on the outside by the positive hydrogen ions, giving rise to a kind of molecular condenser in which the acid is the positive plate, the zinc the negative.

The potential of the zinc thus falls below that of the acid. Hence there will be an electric force tending to drive positively charged ions from the acid into the zinc, counteracting, that is to say, the tendency of the positive zinc ions to dissolve in or combine with the acid. Thus the electric force due to this double layer checks the solution of the zinc, and it can be shewn that the amount of zinc which can be thus dissolved before the action ceases will if the zinc be pure be infinitesimal; we could not expect to detect it by any known

method of analysis.   The state of affairs then is as shewn in
Fig. 181.

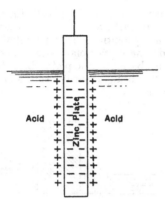

Fig. 181.

The amount of electricity required will depend on the distance
between the two layers of molecules; taking this as 10^{-8} cm.—one
hundred millionth of a centimetre—then the oxidation of $5 \cdot 5 \times 10^{-8}$
grammes of zinc per square centimetre would suffice to give the measured
difference of potential.

If a piece of copper be now placed in the acid the same
kind of action may be supposed to go on, but if we suppose the
force tending to make the copper combine with the acid is less
strong than that which acts in the case of the zinc, a less dense
hydrogen layer will be required to produce sufficient electric
force to balance this force, and the potential difference between
the acid and the copper will be less than that between the
acid and the zinc.   Thus the zinc will be at a lower potential
than the copper.   If the two then be joined by a wire, positive
electricity passes from the copper to the zinc, thus reducing
the potential difference between it and the acid.   This lessens
the electric force which tends to prevent the positive zinc ions
from passing into solution and in consequence more zinc is
dissolved, at the same time the transference of the positive
electricity from the copper to the zinc reduces the potential
of the copper and so produces a further accumulation of the

positively charged hydrogen ions round the copper and thus the current is kept up.

According to this theory of the cell, then, the zinc plate and the copper-wire attached to it are when the circuit is open at the same potential. This potential is lower than that of the acid; the copper plate also is lower in potential than the acid but above the zinc plate, the difference between it and the zinc plate being about 0·8 volt, and this constitutes the E.M.F. of the cell.

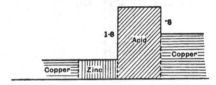

Fig. 182.

This statement of the action of the cell was given in Section 144, and the distribution of potential is shewn in Figs. 182 and 183.

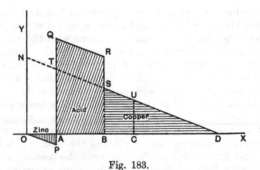

Fig. 183.

If the zinc be not pure the equilibrium condition is not reached; local differences of potential are produced in the zinc about the impurities, local currents are set up and the zinc is continuously dissolved.

192. Volta's Theory. Contact potential. The explanation of the action of the cell given in the last Section is based on the hypothesis, that, except for a small difference depending on temperature, the zinc plate of a cell and the copper-wire attached to it are, when the circuit is open, at the same potential.

There are however a number of experiments which would appear to shew that two metals in contact are at different potentials. Volta was the first to observe this. In some of his experiments he used the condensing electroscope, Fig. 184, described in § 47. For this purpose the plate to which the gold leaves are attached may be made of copper; the upper surface of this plate is covered with a thin layer of shellac or other insulating varnish, the upper plate is of zinc and is carried in an insulating handle. The upper plate is laid on the lower, being separated from it by the varnish, and contact is made temporarily between the backs of the two plates by means of a wire.

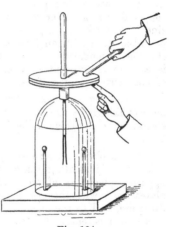

The contact is then broken and the zinc plate removed. When this is done the gold leaves diverge and on testing their charge is found to be negative.

Fig. 184.

We can explain this by supposing that a small potential difference is produced by the contact between the zinc and the copper, the copper being negative. The gold leaves are not sensitive enough to shew this directly, but the two plates constitute a condenser of large capacity and thus a considerable transference of positive electricity takes place from the copper to the zinc; on removing the zinc plate the negative electrification left on the copper distributes itself over the leaves which diverge as indicated. If however the experiment be repeated, but contact be made between the plates by means of a piece of thread soaked in dilute acid which is afterwards removed, no such potential difference shews itself.

These results led Volta to construct the Voltaic pile shewn in Fig. 185, in which a series of discs of copper, zinc and moist flannel are strung together in the order copper, flannel, zinc, copper, flannel, zinc, etc., and in which a potential difference equal to 0·8 multiplied by the number of couples used is produced. According to the theory now under consideration any three adjacent copper, flannel and zinc discs are at one potential, the next copper disc is at lower potential than the zinc in contact with it, but at the same potential as the flannel next to it and the zinc beyond this flannel. Thus there is a drop of 0·8 volt between any pair of consecutive copper discs.

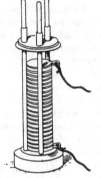

Fig. 185.

These same results are shewn perhaps more strikingly by the aid of the quadrant electrometer and this enables the potential differences involved to be measured.

Thus a plate of zinc and a plate of copper are laid side by side on an insulating stand and connected by wires to the opposite quadrants of an electrometer. On connecting the plates by a wire the electrometer needle is deflected in the direction, and to the amount, which indicate that the zinc is at a higher potential than the copper by 0·8 volt.

The same result may be shewn in an experiment due to Lord Kelvin thus :—

A plate of zinc and a plate of copper are laid side by side in a horizontal position on an insulating stand as in Fig. 186, the two being insulated from each other. An electrometer needle is mounted above them as shewn in the figure ; the axis of the needle being parallel to the adjacent edges of the plates and the needle is electrified, let us suppose positively. On connecting the plates by a piece of wire the needle is deflected, being attracted to the copper plate, thus this plate is negative.

If the copper wire be removed and a drop of dilute acid placed
between the plates no deflexion takes place.

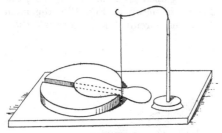

Fig. 186.

Thus it would appear that zinc and copper in contact differ
in potential by about 0·8 volt while when immersed in dilute
acid they are at the same potential. It will be noticed that
this potential difference 0·8 volt is just equal to the E.M.F. of
a zinc copper acid cell. According then to this theory the
action of the cell is as follows. When the zinc and copper
plates are immersed in the acid on open circuit all three are
at the same potential. A copper wire, however, connected to
the zinc plate is at a lower potential than the zinc, at a lower
potential therefore than the copper plate. When then the
other end of the wire is connected to the copper plate positive
electricity flows from the plate into the wire lowering the

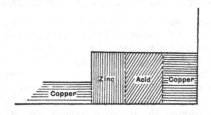

Fig. 187.

potential of the copper plate in the acid, positive electricity
then flows from the zinc plate through the acid to the copper

causing decomposition of the acid and liberating the energy
required to maintain the current. The distribution of potential
when the circuit is open is given in Fig. 187, that when the
current is flowing is shewn in Fig. 188, the main difference
of potential in the circuit takes place at the copper-zinc
junction.

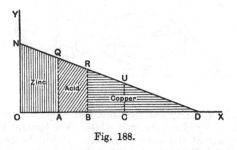

Fig. 188.

193. Chemical and Contact Theories. In order
to reconcile these views we may notice that the contact
potential experiments are ordinarily made in air or with
plates which have been exposed to air. Now the active agent in
the action between the zinc and acid described in Section 190
is probably the oxygen which is also present in air; we may
assume then as probable that the same action tends to go on
between the zinc and the air as between the zinc and the acid
—a potential difference is established between these two.

We suppose—and there are other lines of argument which
lead to the same hypotheses—that an oxygen molecule in air
consists of two atoms, one having a positive charge the other a
negative; these are usually combined but there are a certain
number of free positive and negative ions present. In the
neighbourhood of a plate of zinc negatively charged zinc ions
are escaping from the zinc, these combine with the positive
oxygen ions leaving the negative oxygen ions to form with
the positively charged zinc ions a double sheet over the zinc.
In the same way a potential difference is established between
the copper and the air near it; thus when the copper and zinc
are in contact they are at the same potential, but the air

near the zinc would in this case be at a higher potential than that near the copper. Lines of force pass through the air from the positive film near the zinc to the film near the copper. Hence a positively charged needle placed near is repelled from the zinc to the copper.

194. Contact Experiments in a Vacuum. Attempts have been made to verify this by conducting the contact experiment in a vacuum; they have however failed and calculation shews that, in any vacuum we can produce, there must be a number of oxygen atoms many times larger than is required to produce the double layer over the surface described in Section 190. Hence the result has been that the potential difference observed in a vacuum between copper and zinc is the same as that found in air.

By replacing the oxygen by chlorine a change in the contact difference has been observed; it is extremely difficult however to make sure that this was not due to actual chemical action occurring between the chlorine and the metals.

195. Acid-Metal Contact. When the two metals are connected by acid they are really at different potentials, but the electrometer needle is not deflected, for the air near each of the two metals is at the same potential. The potential of the air near the zinc is as much above that of the zinc as is the potential of the acid, and in the same way

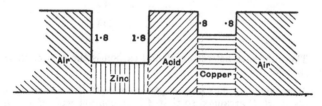

Fig. 189.

the potentials of the acid and of the air near the copper exceed that of the copper by the same amount. The diagram of potential is as shewn in Fig. 189; the air is throughout at the same potential, that of the acid; so far as the metals are concerned there is no field of force and the electrometer needle is not disturbed.

196. Contact Potential. The contact potential difference described as existing between zinc and copper can be observed in a similar manner between any pair of metals and we can arrange the metals as was done by Volta in a series which has the property that the potential in air of any metal in the list—or as we probably ought to express it the potential of the air near any metal in the list—is greater than that of any which follows it.

The following is such a list—zinc lead tin iron copper silver gold platinum carbon—and it should be noted that the order in this list is very nearly the same as the order of the heats of combination between the metals and oxygen.

The potential differences between the consecutive pairs are given in the following Table due to Ayrton and Perry.

Zinc	
	... ... ·210
Lead	
	... ... ·099
Tin	
	... ... ·313
Iron	
	... ... ·146
Copper	
	... ... ·238
Platinum	
	... ... ·113
Carbon	

The difference between zinc and carbon is the sum of these numbers or 1.

Moreover it is shewn by experiments that if we have three metals A, B, C arranged so that A is in contact with B and B with C the difference of potential between A and C is exactly what we observe when A and C are put into connexion directly, so that if we connect to C a second piece A' of the metal A, then A' is at the same potential as A; hence, if A and A' be joined we do not get a current in the circuit.

Let us for example suppose that there is a rise of ·5 volt between A and B and a further rise of ·8 between B and C so that C is 1·3 volts above A, it will be found that there is a fall of 1·3 volts between C and A'; thus A and A' are at the same potential, if they be joined the total electromotive force round the circuit at the three junctions is ·5 + ·8 − 1·3 or zero.

On the chemical theory of course the metals when in contact are all at the same potentials; the differences occur between the portions of air in the immediate neighbourhood of the respective conductors. The distribution of potential would be as in Fig. 190 where the lower line gives the potentials of the metals, the upper that of the air in contact with them.

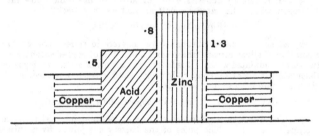

Fig. 190.

It has been tacitly assumed throughout the above that the conductors are throughout at the same temperature. If in a circuit of two metals one junction is maintained at a higher temperature than the other, then as we have already seen a current flows round the circuit. See Section 182.

EXAMPLES ON VOLTAIC ELECTRICITY.

1. If the electrochemical equivalent of hydrogen is $1.03 - 10^{-4}$ what are the electrochemical equivalents of (1) oxygen, (2) iron, first in ferrous, second in ferric salts?

(Atomic wt. of iron = 56, ferrous iron is divalent, ferric iron is trivalent.)

2. If a current of 4.8 amperes will decompose 8 grams of water in 5 hours, what is the electrochemical equivalent of hydrogen?

3. If the electrochemical equivalent of hydrogen is .0001 gr., how much copper will be deposited in one hour by a current of one ampere?

(Atomic wt. of copper = 63.4.)

4. If 1 ampere deposits 4 gr. of silver in one hour, how much hydrogen will be liberated by half an ampere in a week?

(Equivalent wt. of silver = 108.)

5. A current passes through 3 voltameters in series; one contains a solution of silver nitrate, the second a solution of copper sulphate, and the third acidulated water; it is found that 2.7 gr. of silver are deposited. Calculate the mass of copper, hydrogen and oxygen liberated, having given the following atomic weights.

(Hydrogen = 1, oxygen = 16, copper = 63.4, silver = 108.)

6. The E.M.F. of a given battery of cells of total internal resistance 5 ohms is 10 volts. The poles of the battery are joined by a wire of resistance 25 ohms. Find in C.G.S. measure the current produced, and the difference of potential between the poles.

7. The E.M.F. of a battery on open circuit is 4.8 volts; when producing a current of 1.5 amperes through a wire the difference of potential sinks to 3 volts; find the resistance of the wire and of the battery.

8. The E.M.F. of a battery on open circuit is 10 volts. When producing a current of 5 amperes the difference of potential between its poles is 8 volts, find its resistance.

9. The E.M.F. of a battery is 15 volts and its total internal resistance is 6 ohms. The poles of the battery are joined by a wire of resistance 30 ohms. Find the current produced and the difference of potential between the poles.

10. A cell gives a current of 1 ampere when its terminals are joined by a wire of no appreciable resistance, and $\frac{1}{3}$ ampere when joined by a wire of 2 ohms resistance; find its E.M.F. and its resistance.

11. A cell sends a current of 1 ampere through an external resistance of 1 ohm, and 2 amperes through an external resistance of $\frac{1}{4}$ ohm. Find its E.M.F. and its resistance.

12. Account for the fact that a Leclanché cell will send a greater current through a long thin wire than a Daniell's cell, while if the wire is short and thick the reverse is the case.

13. A cell of E. M. F. 1 volt sends a current of $\frac{1}{2}$ an ampere through a galvanometer whose resistance is 1 ohm. What current would be sent by 3 such cells in series through the same galvanometer?

14. You have 12 Grove cells, each giving an E. M. F. of 1·9 volts and having a resistance of $\frac{1}{4}$ ohm. How would you couple them so as to get the greatest possible current through a resistance of $\frac{1}{4}$ ohm?

15. Two equal cells when connected in series through a given wire produce a current of 10 amperes, while when connected in multiple arc through the same wire the current is 7·5 amperes. Shew that the resistance of the wire is $2\frac{1}{2}$ times that of either cell.

16. How must a battery of 10 cells, each having a resistance of 2 ohms, be arranged so as to give the largest current through an external resistance of 6 ohms?

17. Three cells each having an E. M. F. of 1 volt and an internal resistance of 2 ohms are available; how would you arrange them so as to obtain the greatest current through a circuit of 2 ohms resistance?

18. Ten Daniell's cells, each having an E. M. F. of 1 volt and an internal resistance of 2 ohms, are arranged in 2 parallel groups, the 5 cells of each group being placed in series. Calculate the current which they will send through a galvanometer having a resistance of 30 ohms, with its terminals connected through a coil whose resistance is 6 ohms.

19. Ten voltaic cells, each of E. M. F. 1·75 volts and resistance ·75 ohm, are joined in series and the circuit completed by a wire of resistance 12·5 ohms. Find (1) the strength of the current, (2) the quantity of electricity that passes any section of the circuit per minute, (3) the difference of potential at the terminals of the battery.

20. You are given 48 cells each of E. M. F. 1·8 volts and resistance ·3 ohm. How would you arrange them to produce the greatest current in a circuit of 5 ohms resistance?

21. A Daniell's cell has E. M. F. 1·1 volts and internal resistance ·5 ohm. Its terminals are joined by 2 wires arranged in parallel of resistances 1 and 1·5 ohms respectively. Find the current in each wire and in the cell.

22. You are given a battery of E. M. F. 5 volts and resistance 2 ohms, a coil of wire of resistance 8 ohms and a galvanometer of resistance 5 ohms. Calculate the current (a) through the cell and (b) through the galvanometer when (1) all three are in series, (2) the poles of the battery are connected to the galvanometer and the coil in multiple arc.

23. Two electrolytic cells each containing copper sulphate the resistance of which is very high compared with all other resistances in the circuit are placed first in series, and secondly in multiple arc; compare the total quantities of salt decomposed in the two cases.

24. The poles of a Daniell's cell (E. M. F. 1·08 volts, resistance ·5 ohm) are connected in multiple arc by two wires ACB and ADB; the resistances of ADB and the part AC are each 1 ohm, what must be the resistance of the part CB in order that the difference of potential between A and C may be ·01 volt?

25. Two points A and B are joined by two conductors ADB and ACB. The one ADB has a resistance of 1 ohm, the other ACB is 98 ohms between A and C and 1 ohm between C and B. If a current of 1 ampere enters the system at A and leaves it at B, find the difference in potential between C and B.

26. A current passes through a water voltameter A, and then divides and passes partly through a copper voltameter B and partly through a silver voltameter C, B and C being arranged in parallel. If 1 gr. of copper is deposited in B and 2 gr. of silver in C, how many gr. of hydrogen will be set free in A?

(Atomic wt. of copper = 63·4, of silver = 108.)

27. Two wires of resistances 50 and 10 ohms respectively connect two points. What is the effective resistance of these two wires when in multiple arc?

28. If two conductors of 3 ohms and 4 ohms resistance respectively are joined in parallel what is their combined resistance?

29. Enumerate all the resistances that can be obtained from 3 coils of resistances 2, 4 and 6 ohms respectively by the various ways in which they may be connected, all three coils being always in use.

30. A galvanometer having a resistance of 5000 ohms is shunted with 100 ohms. A certain deflexion of the galvanometer is obtained with a battery of constant E. M. F. when the resistance of the rest of the circuit is 2000 ohms. What additional resistance must be inserted to produce the same deflexion when the shunt is removed?

31. A battery is connected up by thick wires to a galvanometer and the current is observed. On shunting the galvanometer with $\frac{1}{71}$ of its own resistance the current is halved; shew that the resistance of the galvanometer is 20 times that of the battery.

32. If a battery of very low internal resistance R is connected with the terminals of a galvanometer, the deflexion is almost unaltered when the instrument is shunted; but if the internal resistance of the battery is high, the alteration is very considerable. Explain this.

33. A battery is connected by short thick wires to a galvanometer and the deflexion noted. The galvanometer is then shunted with $\frac{1}{8}$ of its own resistance, and on again connecting with the battery the current through the galvanometer is observed to have half its former value. Shew that the resistance of the battery is half that of the galvanometer.

34. When 2 coils of resistances 10 ohms and 5 ohms respectively are connected up in series with a galvanometer and a battery of negligible resistance the current indicated is ·2 ampere; when the coils are in parallel the current is ·35 ampere. Calculate the resistance of the galvanometer.

35. A reflecting galvanometer has a resistance of 100 ohms and is shunted with $\frac{1}{10}$ ohm; a battery of very low internal resistance and E.M.F. of 2 volts is put in series with it and 10,000 ohms. The scale deflexion observed is 150 divisions; find the sensibility.

36. A galvanometer of 100 ohms resistance is placed in series with a resistance box of 45 ohms and with a battery whose E.M.F. is 1·5 volts and resistance 5 ohms, and the deflexion is observed. The resistance box is then short circuited and the same deflexion as before is produced by shunting the galvanometer. Find the resistance of the shunt.

37. The resistance of a piece of wire 1 mm. radius and 15·7 metres long is 1 ohm, find its specific resistance.

$$(\pi = 3·14.)$$

38. The specific gravity of silver is 10·5. The resistance of a silver wire 100 cm. long and 1 gr. in weight is ·1689 ohm. Shew that the specific resistance of silver is 1609 using absolute units.

39. A wire is stretched uniformly until its length is doubled. Compare its resistance before and after stretching.

40. A uniform copper wire, whose resistance is 12 ohms, is bent into the form of a square and the ends soldered; the poles of a battery, whose resistance is 3 ohms, are joined at two opposite corners A and C of the square; in what ratio will the strength of the current flowing along each side of the square be altered by joining A and C by a straight piece of the same copper wire?

41. A battery when joined to a tangent galvanometer of 10 ohms resistance gives a deflexion of 60°. If a resistance of 20 ohms is inserted in the circuit the deflexion falls to 45°. What is the resistance of the battery?

42. Two currents passed in turn round a tangent galvanometer produce deflexions of 30° and 60° respectively. Compare the strengths of the two currents.

43. A tangent galvanometer has 10 turns of wire wound in a groove of radius 20 cm.; what current (expressed in amperes) will give it a deflexion of 30°?

$$(H = ·18 \text{ c.g.s. unit.})$$

44. A wire is coiled into a circle of 10 turns and used as the coil of a tangent galvanometer. On passing a current of 1 ampere the deflexion is 45°. Find the radius of the circle.

$$(H = ·18 \text{ c.g.s. unit.})$$

45. A current of 5 amperes flows in a circular wire of 10 cm. radius. How many turns of wire are there in the coil if the strength of the magnetic field at the centre is 1 dyne?

46. The coil of a tangent galvanometer is 10 cm. in radius; how many turns of wire must be wound on if a current of ·01 ampere is to produce a deflexion of about 45°?

$(H = ·18$ c.g.s. unit.)

47. Two Daniell's cells give equal deflexions on a quadrant electrometer, but quite different deflexions when connected with a low resistance galvanometer. What do you suppose is the cause of the difference?

48. A tangent galvanometer has a coil of 10 turns of wire with a radius of 10 cm. What mass of copper will be deposited from a cupric salt in half an hour by a current which deflects the needle through 45°?

$(H = ·18$ c.g.s. unit. $\pi = 3·14.$ 1 ampere deposits ·000328 gr. of copper in one second.)

49. Two copper plates are immersed in a solution of copper sulphate and a current passed through them and a tangent galvanometer. The deflexion of the galvanometer is 45°, and after an hour it is found that 216 milligrammes of copper have been deposited on one plate; having given that a current of 1 ampere deposits 19·8 milligrammes per minute, deduce the reduction factor of the galvanometer.

50. An arrangement of resistance keys and connecting wires is made for the purpose of determining an electrical resistance. If when the galvanometer circuit is made, the battery key being open, a deflexion of the galvanometer is produced, by what experiments would you determine whether the deflexion is due to (1) leakage through the battery key or (2) an E.M.F. in the circuit independent of the battery?

51. The resistance of a coil of copper wire is determined by a Wheatstone's Bridge box when the temperature of the air is 20° C. and is found to have the value 20·25 ohms. Calculate its true value at 0° C., the coils of the box being of german silver and correct at 15° C.

Coefficient of increase of resistance per 1° C. for copper is ·0038
,,   ,,   ,,   ,,   german silver is ·0004.

52. A careless observer in setting up his apparatus for the measurement of the resistance of a coil by means of the Wheatstone Bridge neglects to clean the ends of the wires by which the connexions are made between the coil, galvanometer and battery and the rest of the apparatus. Consider the effect of the neglect of these precautions upon (1) the result, (2) the sensitiveness of the method.

53. A wire whose resistance was to be determined was placed in a Wheatstone's Bridge, the fixed ratio arms of which were 10 and 100 ohms respectively; balance was obtained when the adjustable coils were arranged to give a resistance of 467 ohms. What was the value of the resistance of the coil under examination?

54. If a stretched and graduated wire whose resistance is 5 ohms is connected to a battery whose E.M.F. is 3·1 volts and internal resistance 1·2 ohms, what is the limit of the E.M.F.'s which can be compared by means of it, using the potentiometer method?

55. A fine wire is placed in 100 gr. of water in a light copper calorimeter and 2 amperes are passed through it, an E.M.F. of 14 volts being maintained between the ends of the wire. Calculate the rise of temperature in 10 minutes.

$$(J = 4·2 \times 10^7 \text{ ergs.})$$

56. Through a coil of wire which is immersed in water in a small calorimeter a current of 2 amperes is passed; the calorimeter contains 200 gr. of water at a temperature of 15° C.; at the end of 5 minutes it is found that the temperature has risen to 18° C., find the value of the electrical resistance of the coil.

57. A platinum wire has a resistance of half an ohm. You are provided with 8 Daniell's cells each of 1 ohm resistance. How would you arrange them in order to make the wire as hot as possible?

58. A battery of E.M.F. 6 volts and resistance 2 ohms is connected to two wires in parallel of 1 and 3 ohms respectively. Find the current and the heat developed per second in each branch.

59. The E.M.F. of a constant battery is 3 volts; its internal resistance is 1 ohm. The battery sends its current in the first place through a copper wire of 3 ohms' resistance. The wire is then removed and an electrolytic cell containing dilute sulphuric acid, also of 3 ohms' resistance, substituted in its place. If it require an E.M.F. of 1·1 volts to electrolyse dilute sulphuric acid, compare the heat generated per minute in the acid with that generated in the wire.

60. A current is passed through a thin wire enclosed in a calorimeter, and through a copper voltameter arranged in series with the wire. Calculate the resistance of the wire from the following data :—

Time for which current is passed .... = 20 minutes.
Weight of copper deposited .... = ·3 gram.
    ,,   ,, water in the calorimeter .... = 100 grams.
Water equivalent of the calorimeter .... = 10 grams.
Rise of temperature corrected for radiation losses... = 20° C.
Mechanical equivalent of heat .... = 4·2 × 10⁷ ergs.
Electrochemical equivalent of copper .... = ·0033 gram per C.G.S.
                                                         unit of current.

61. What will be the ratio of the currents which will keep two wires of the same material heated to the same temperature if the radius of one wire is double that of the other?

62. If the loss of heat from a wire per unit area of surface by radiation and convection be proportional to the excess of its temperature above that of the surrounding air, shew that, if the same current is sent through wires of the same material, the elevation of temperature is inversely proportional to the cube of the radius of the wire.

63. A copper wire ·02 cm. in diameter, carrying a current of 1 ampere, is found to reach a steady maximum temperature of 100° C. Taking the specific resistance of copper at 100° C. as $2 \cdot 1 \times 10^{-6}$ ohms per centimetre cube, and the value of J as $4 \cdot 2 \times 10^7$, calculate how many units of heat are emitted per second by 1 sq. cm. of copper surface at 100° C.

64. A battery of E.M.F. E and internal resistance B drives a current through a wire of resistance R. Shew that the heat produced in the wire in a given time is a maximum when $R = B$.

65. Find the horse-power required to light 75 incandescent lamps each taking $\frac{1}{2}$ ampere and requiring 100 volts at its terminals. If one of these lamps were immersed in 2 litres of water and then lighted, how fast would the temperature of the water rise?

(A horse-power $= 746$ watts $= 746 \times 10^7$ ergs per sec.)

66. Power to the extent of 100,000 watts has to be carried to a distance of 5000 metres with a loss not exceeding 5 %. Compare the cost of the copper mains if the current has a voltage of 100 with their cost if the voltage is raised to 2000.

67. A difference of potential of 100 volts is maintained between the terminals of a dynamo machine supplying an installation of 100 lamps; the current in each lamp is ·65 of an ampere and the resistance of the leads to each lamp is 1 ohm. Calculate the amount of energy supplied by the machine per hour and the amount wasted in the leads.

68. A pair of copper bars are required to transmit 200 amperes from a dynamo to a motor at 1000 metres distance. The E.M.F. at the dynamo is 100 volts and the E.M.F. at the motor is not to be less than 96 volts. If the specific resistance of copper is $1 \cdot 6 \times 10^{-6}$ ohms per centimetre cube, calculate the least section each bar may have.

69. A current is passed across the junction of two metals, and it is found that heat is evolved at this junction. How would you distinguish between the heat generated by overcoming the electrical resistance at the junction and the heat due to the Peltier effect?

70. A strong current is sent for a short time across the junction of a bar of antimony and a bar of bismuth. A galvanometer is then placed in the circuit instead of the battery and it is found that the galvanometer needle is deflected. Explain this phenomenon.

CHAPTER XX.

ELECTROMAGNETISM.

197. Magnetic Force due to a Current. We have already seen (§ 112) that the space in the neighbourhood of a current is traversed by lines of magnetic force—it is a field of force—and we have discussed in the case of a galvanometer how this fact may be used to measure a current. Reference has also been made to the magnetisation of iron by a current. These facts are of the greatest importance in the modern applications of electricity and we must now consider them more closely.

We can trace experimentally in various ways the lines of force due to a current in a coil of wire; thus take a circular coil and fix it so that its plane is vertical; adjust a horizontal sheet of cardboard as shewn in Fig. 191 so that the axis of the coil may lie in the sheet of card and sprinkle iron filings over the card. On passing a current through the coil and tapping the card the iron filings arrange themselves along the lines of magnetic force in the plane of the coil.

Or again if a small compass needle be placed on the card the needle will set itself when the current passes so as to be tangential to the line of force through its centre; the direction of the line of force will pass through the needle from its south pole to its north.

The lines of force start from one face of the coil and travelling outwards gradually curve round and enter the other face of the coil. In this they resemble the lines of force from a magnet which start outwards from the north pole and passing round enter again at the south pole.

21—2

It must be remembered in these and similar observations that the lines of force observed are due to the resultant action of the current in the coil and the earth's magnetism. In Fig. 191 the lines are drawn as undisturbed by the earth's force.

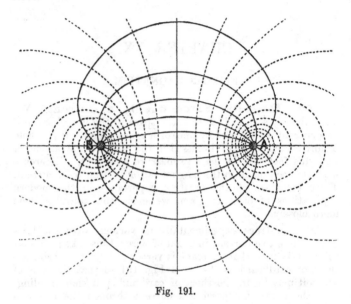

Fig. 191.

We may shew as the result both of theory and experiment that if the coil be small the distribution of magnetic force at some distance away is the same as that due to a small magnet placed with its centre at the centre of the coil and its axis at right angles to the plane of the coil. Moreover if the total area of the windings of the coil be A square centimetres and the strength of the current in the wire be i c.g.s. units, then the magnetic moment of the equivalent magnet can be shewn to be Ai.

In estimating A the number of turns of wire on the coil has to be taken into account; it is the sum of the areas of the individual turns; thus if the coil consist of n turns each of area a the value of A is na.

198.   Magnetic Shells.   If however the coil cannot be treated as very small we cannot represent its action by that of a single small magnet, we have to suppose the area of the coil divided into a number of small equal elements or parts and place at the centre of each element a small magnet with its axis at right angles to the plane of the coil.   These small magnets are all equal and their north poles all point in the same direction.   Moreover if there be n of them and if A be the total area of the coil and i the current, then the magnetic moment of each is Ai/n.

Such a distribution of magnets constitute if they are placed quite close together what is known as a magnetic shell. We may look upon them as forming a sheet of iron or magnetic material bounded by the coil; one side of this sheet will be coated with north poles, the other side with south poles.   The distribution of magnetic force due to the coil will then be the same as that due to the shell.

It is important to determine readily which face of the coil is to be treated as the north-pointing or positive face.   The right-handed screw rule given in Section 112 enables us to do this.   Imagine the right hand placed on the wire as in Fig. 192 so that the current is passing from the wrist out at the fingers and let the extended thumb be within the area formed by the coil; then the back of the hand represents the north pole ends of the equivalent magnets—the north face of the coil we may call it.   This is obvious if we remember that a north pole placed at the end of the thumb would move round the current in the same direction as the hand must be turned to screw in a right-handed screw whose point coincides with the fingers and is to be moved

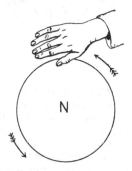

Fig. 192.

in the direction in which the current is travelling.   To secure this the direction of motion of the thumb must start outwards from the back of the hand; thus the lines of force issue from the back of the hand placed in the position described and enter at the front. The back of the hand is the north face of the coil.

Another rule may be given which is sometimes more convenient.

Place a watch with its face towards the observer and coincident with the plane of the coil.

If the current circulates in the coil in the same direction as the hands of the watch move then the south face of the coil is towards the observer, Fig. 193 ; if the direction of the current is opposite to that of the motion of the hands the north face is towards the observer, Fig. 194.

S   N

Fig. 193.   Fig. 194.

199. Action of a Magnet on a Current. If a coil of wire carrying a current behaves like a magnetic shell and produces a magnetic field in its neighbourhood, then if another coil or a magnet be brought near there will be magnetic force between the two. It was by studying experimentally the forces between a magnet and a coil carrying a current that the equivalence of a current and a magnetic shell was established.

Thus let a coil consisting of a single turn of wire be supported so that it can rotate readily about a vertical axis. This is done in Ampère's stand by balancing the coil on a steel point which rests in a cup at the top of a vertical axis. One end of the coil is connected to this point.   A second annular cup surrounds the first but is insulated from it and both cups contain mercury.   The second end of the coil is connected to a needle which dips into the mercury in the annular cup and maintains electric contact between it and the cup as the coil rotates round the central pivot; if then one pole of the battery be connected to the outer cup and the other pole to the central cup the current passes through the coil which is free to move round a vertical axis.

The arrangement is shewn in Fig. 195, where A is the inner cup, G the outer.

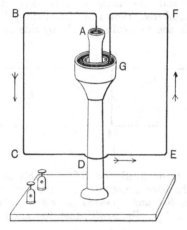

Fig. 195.

$ABCDEFG$ is the wire balancing on a pivot at A.   There is a bend in the wire at D to allow it to clear the support. If a current be passed through the coil from A to G it will be found that the face $ABCEFG$ becomes a north face; the coil sets itself with its plane pointing east and west; the upper surface of the paper will become a north pole.   This can be also tested by bringing the north pole of a magnet near the coil, the face in question will be repelled, the opposite face attracted.

This action of course is the converse of that made use of in the ordinary galvanometer already described in which the coil is fixed and the magnet moves.

200. Electromagnetic Action between two Currents. Again, if a second coil carrying a current is brought near the suspended coil, forces of attraction or repulsion come into play. The direction of these forces may be best determined by considering the action between the two magnets to which they are equivalent; the moveable coil will tend to set itself so that the axes of the magnets to which it is equivalent lie as nearly as may be along the lines of force due to the second coil; the direction of the axis of a magnet, it must be remembered, runs from its south pole to its north pole. Thus if the two coils be side by side as in Fig. 196 a

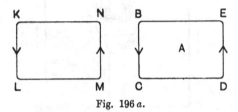

Fig. 196 a.

the coil KM being fixed and BD moveable and the current circulate in the same direction round the two so that it runs downwards in BC, and upwards in MN, a side of the fixed coil adjacent and parallel to BC, then the north faces of both

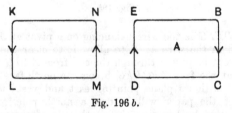

Fig. 196 b.

coils are on the upper side of the paper, the lines of force from the fixed coil KM pass through the moveable coil BD

downwards, that is from north to south; thus the moveable coil will tend to set itself in the reverse position so that the lines of force may traverse it from south to north, hence ED becomes adjacent to NM and the currents in the two parallel wires run in the same direction as in Fig. 196 b.

Thus two parallel wires carrying currents repel each other when the currents are in opposite directions, they attract when the currents run in the same direction.

Again let the fixed coil be placed below the moveable one, the planes of both being vertical but not coincident and the sides of the coils horizontal and vertical.

Then looking down on the coils their projections in a horizontal plane would be EOB and LOM, Fig. 197 a, while the current runs from L to M and from E to B, where LM is the upper side of the fixed circuit, EB the lower side of the moveable one.

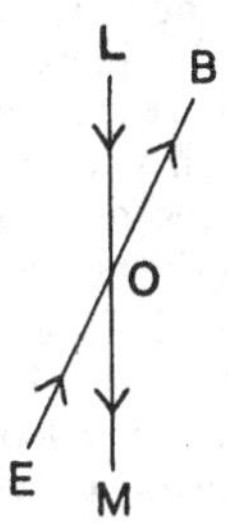

The north faces of both circuits are to the right. The lines therefore from the upper fixed circuit pass through the lower moveable circuit from north to south, thus the lower circuit tends to reverse its position and to set itself so that the lines traverse it from south to north, the two wires tend to

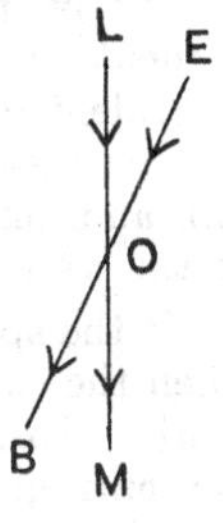

Fig. 197 a.   Fig. 197 b.

become parallel while the currents in the two parallel wires run in the same direction, as in Fig. 197 b.

Thus when the currents run in two adjacent wires as in Fig. 197 a the acute angle between the wires is increased by the action between the currents, if the current were reversed in one wire, say LM, the acute angle would be decreased.

In the first case, limiting ourselves to the portions of the wire below the point O, the current in one wire is running towards the angle, in the other wire it is running from the angle; in the second case with BE reversed the current in both wires is running from the angle.

In all these cases the moveable circuit tends to set itself

so that the number of lines of force which traverse it from south to north should be as great as possible.

201. Solenoid. By making the coil in the form of a spiral having a considerable number of turns as in Fig. 198

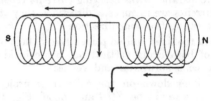

Fig. 198.

instead of the single turn shewn in Fig. 195 its magnetic moment is greatly increased.   Each spire of the coil is equivalent to a short magnet of small moment, or more strictly to a magnetic shell with its axis lying along the axis of the coil, and this series of magnetic shells constitutes a magnet of the same length as the axis of the spiral.

If the spiral be balanced at its centre on the ampere stand, when the current passes, it sets itself with its axis north and south.   To an observer looking at the north pole of the coil the current appears to circulate as in Fig. 199 a, in the opposite direction to the motion of the hands of a watch, to an observer looking at the south pole the direction of the

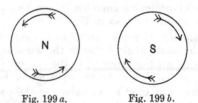

Fig. 199 a.   Fig. 199 b.

current appears to circulate as in Fig. 199 b, in the same direction as the hands of the watch move.

If the spiral be sufficiently long the magnetic force within it is very approximately uniform, the lines of force are parallel to the axis of the spiral, and it can be shewn that at a sufficient distance from the ends this force is equal to $4n\pi i/l$ where i is the current in c.g.s. units, n the number of turns, and l the length of the axis of the spiral, or if N denote the number of turns per unit length of the solenoid then $N = n/l$ and the value of the force[1] is $4\pi Ni$.

202. Astatic Coil. If the coil be as in Fig. 195 the earth's magnetic field acts on it and may render it unsuitable for delicate observations, it can however be rendered astatic by bending the wire as in Fig. 200.

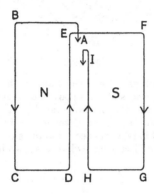

Fig. 200.

If the current flow as indicated by the arrows the face of $ABCDE$ turned towards the reader is a north pole, that of $EFGHI$ a south pole, the earth acts equally and oppositely on the two and the system is astatic.

203. Electromagnetic Forces. We may regard the above results somewhat differently thus. We have seen that two parallel wires carrying currents in the same direction attract. Let us suppose the wires to be perpendicular to the paper and to cut it in A and B, Fig. 201, and let the currents flow downwards. Then if the wires be sufficiently long

[1] If the current is measured in amperes the magnetising force is $4\pi Ni/10$.

compared with the distance AB, the lines of magnetic force round them are circles, as shewn by the dotted lines in the figure.

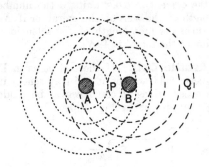

Fig. 201.

At a point such as P between A and B the magnetic forces due to the two currents act in opposition, at a point such as Q the directions of the two forces coincide.   Thus to the right of B the magnetic field is strengthened by the presence of A, between A and B it is weakened; the lines of force due to the combined system will run somewhat as shewn in the Fig. 201 a.

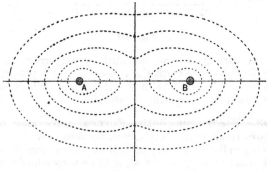

Fig. 201 a.

Now we have seen in the case of the electrostatic field that many of the phenomena can be explained by supposing, (1) that the lines of force are attached to the conductors in which they terminate, and (2) that there exists in the field (i) a tension acting along the lines of force tending to shorten them, and (ii) a pressure at right angles to the lines of force tending to press them apart. Clearly the result of such action in the magnetic field would be to bring the wires A and B together, to cause each to move in a direction at right angles to its own length and to the lines of force in the field.

We shall find that the hypothesis that a force is impressed on each unit of length of a wire carrying a current in a magnetic field which is proportional to the product of the strength of the field and the current, and perpendicular to both the wire and the lines of magnetic force, and acts in the direction in which the strength of the field decreases, will enable us to account for the attractions and repulsions observed, and it is a consequence of theory that such a force should exist.

Thus consider a vertical wire in a uniform horizontal magnetic field H acting from south to north. Let the wire cut the paper in Fig. 202, and let the current flow downwards.

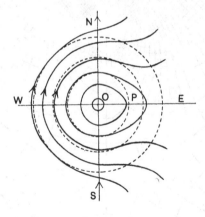

Fig. 202.

The lines of force due to the current i are the dotted circles.

Let *EOW* be the east and west line through *O*.

At a point *P* between *O* and *E* the force due to the current is opposite to that due to the field, to the west of *O* the reverse is the case, thus the field is strengthened by the

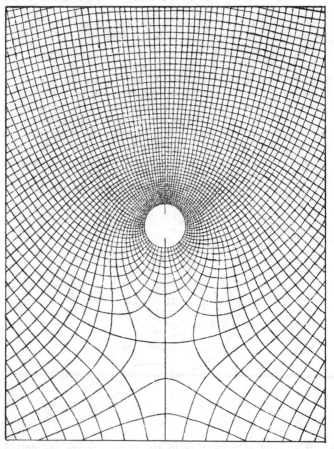

Fig. 203.

current to the west of O, weakened to the east. According to the law the conductor is driven from west to east by the electromagnetic force with a force per unit length equal to Hi. If the current be upwards the magnetic force is from east to west.

Fig. 203 gives a more complete drawing of the same.

The magnetic force acts horizontally from right to left and the current runs vertically downwards.

From these considerations we can get a rule connecting together the directions of the current, the magnetic field, and the electromagnetic force.

RULE. *If a right-handed screw be placed with its axis perpendicular to the plane containing the current and the lines of magnetic force, and turned so that the direction of rotation is from the direction of the current to that of the field, then the point of the screw moves in the direction of the electromagnetic force.*

The various directions are shewn in Fig. 204.

In all cases the force per unit length on the conductor is equal to the product of the current and the component of the field resolved at right angles to the current. If H be the strength of the field, a the angle between it and the current i, then the component in question is $H \sin a$ and the electromagnetic force is $Hi \sin a$. This law was first established by Ampère by direct experiment, the experiments however are difficult and not conclusive; the real proof of the law lies in the fact that it is made use of to calculate the theoretical consequences of many complex experiments and these theoretical consequences are found to agree with experimental results.

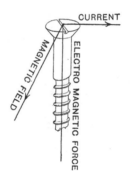

Fig. 204.

Another form of the rule, due to Professor Fleming, may sometimes be found more convenient. Extend the first finger of the left hand horizontally, as in Fig. 205, point the second finger downwards, and the thumb to the right. Then keeping them in these relative positions, place the hand so that the first finger may be parallel to the lines of magnetic force, and the second finger parallel to the current, the thumb will then give the direction of the electromagnetic force.

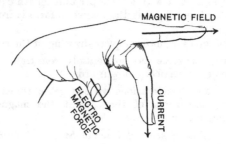

Fig. 205.

204. Electromagnetic Action between Conductors. We may use these results to explain the attractions and repulsions described in Section 200.

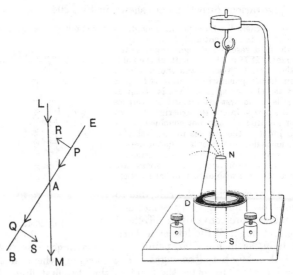

Fig. 206.   Fig. 207.

For let us suppose the wire LAM carrying a current from L to M to be fixed, and consider another wire EAB, Fig. 206, in which a current flows from E to B. Then the magnetic field in which EAB is placed is that due to the current in LM. At a point such as P the magnetic force is downwards, hence the electromagnetic force is along PR perpendicular to EAB; while at Q it is along QS, also perpendicular to EAB. In consequence of these forces EAB tends to set itself parallel to LAM.

Other cases may be treated in the same way.

205. Electromagnetic action. Equilibrium condition. ˙ It follows as a consequence of the above law of electromagnetic force that a circuit, which is free to move in a magnetic field, when carrying a current, sets itself in such a position that the number of lines of magnetic induction which pass through it from its south face to its north may be as large as possible.

A loop of wire suspended freely on the ampere stand, Section 199, sets itself with its plane east and west, and its south face pointing south, the axis of the spiral shewn in Fig. 198 above will point north and south.

If a current be passed through a perfectly flexible conductor which is supported in a plane at right angles to a uniform magnetic field, the conductor takes the form of a circle, for in this way it includes in its area the maximum number of lines of force. This experiment may be approximately realized by the use of a thin strip of gold-leaf or tin-foil through which a current is passed.

206. Motion of a conductor in a magnetic field. The electromagnetic force on a conductor carrying a current may be shewn in various ways.

Thus in Fig. 207, NS is a strong cylindrical magnet fixed in a vertical position. The lower end D of a wire CD is supported so that it can turn freely about a vertical axis through the magnet. The upper end can turn about C, a point in the axis SN produced, and the wire rests in an oblique position. The end D dips into an annular mercury

cup, and from thence a current can be passed along the wire.
Lines of magnetic force issue from the magnet and cut the
wire. If a current be passed through the wire it will be
set into continuous rotation, the direction of which will
depend on that of the current. If the current be downwards
from C to D, the end D of the wire will move from the
observer.

Barlow's wheel, Fig. 208, affords another example of the
motion of a conductor carrying a current in a magnetic field.

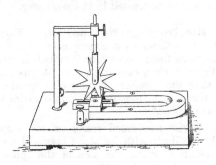

Fig. 208.

A star-shaped disc can rotate easily about a horizontal
axis. The points of the star as the disc rotates dip into a
little mercury contained in a hollow in the stand of the
instrument, and a current passes through the disc between
the axis and this mercury.

A horseshoe magnet is placed so that its poles are on
either side of the mercury cup; thus the disc can rotate
between the poles of the magnet, the lines of magnetic force
traverse the disc in a direction at right angles to the current,
and the result is that the disc is set into rotation, which
continues so long as the current is maintained.

The rotation of a current about a current can be illustrated
by the apparatus shewn in Fig. 209. A wire frame is pivoted
on to the top of a pillar rising from the centre of a circular
trough which contains copper sulphate or some conducting
liquid.

The trough is surrounded by a horizontal coil of wire, and the connexions are such that a current after traversing this coil can pass up the fixed stem and then along the two branches of the wire *BC*, *BD*, and away again through the liquid in the trough.

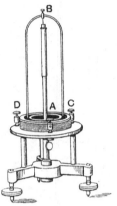

The fixed coil produces a magnetic field whose lines of force cut the moveable conductors *BC* and *BD*. The electromagnetic forces on these two conductors both act at right angles to the paper but in opposite directions, thus the moveable frame rotates about the central pivot.

207. Rotation of a magnet about a current. If a magnet *NS*, Fig. 210, be supported as in the figure so that it can rotate about a vertical wire in which a current flows downwards remaining always parallel to the

Fig. 209.

wire, the force on the south pole tends to bring it forward from the paper, that on the north pole tends to depress it below the paper, and the movements of these forces about the axis being equal the magnet does not rotate.

But suppose the arrangement is as in Fig. 211, where two magnets are shewn pivoted in a mercury cup at *A*. At *B* is a wire dipping into an annular cup.

The current enters through the central vertical rod, passes to the annular cup by the wire attached to the magnets, and down the stand to the binding screw. The north poles of the magnets are above the axial current, and hence the force on them is less than on the south poles which are close to the central conductor. Thus rotation takes place.

The results we have just been considering as to the motion of a circuit carrying a conductor in a magnetic field have important practical applications. The action of the electric motor depends entirely on them, and Barlow's wheel is in

a sense the parent of the modern applications of electricity to
the supply of motive power.

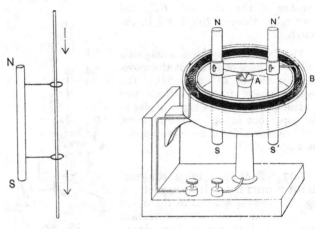

Fig. 210.   Fig. 211.

The principles of some simple form of motors are described
briefly in Section 247.

CHAPTER XXI.

MAGNETISATION OF IRON.

208. Electromagnets. The fact that iron can be magnetised by a current was discovered by Arago. Since iron is magnetised by being brought into a magnetic field and a current flowing in a wire produces such a field it is clear that the magnetisation will take place. Arago shewed that a spiral of copper wire in which a current is passing if dipped into iron filings became coated with the filings like a magnet. If again a piece of copper wire covered with some insulating material be wound in a spiral on a tube as in Fig. 212, and

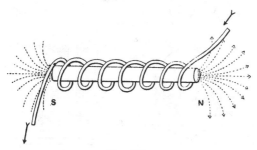

Fig. 212.

a current be passed through the wire, then the interior of the tube becomes a field of force and a piece of steel placed in the tube and shaken while there becomes permanently magnetised. If the tube be filled with a core of soft iron the lines of magnetic force are concentrated through the iron which while the current lasts becomes a powerful magnet; on breaking the current much of the magnetism of the iron disappears.

The laws we have already investigated enable us to tell which end of the iron core will be a north pole. For let the current appear to an observer looking at the spiral to circulate in the direction shewn in Fig. 213 a. The lines of force in

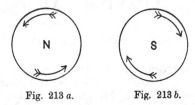

Fig. 213 a.   Fig. 213 b.

the interior of the spiral pass through the paper from below upwards leaving the paper on its upper side; thus the end

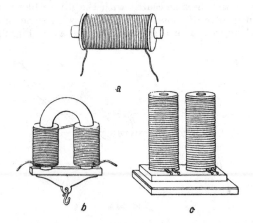

Fig. 214.

looked at is a north pole: if however the direction of the current be as in Fig. 213 b, the lines of force enter the paper from above; the end is a south pole. This we have seen already in Section 201. Electromagnets take various forms as shewn in Fig. 214 *a*, *b* and *c*. In all cases we have the magnetising coil or coils wound on a continuous core of soft iron; the core should be continuous, for the object is to confine within the iron as many magnetic lines as possible, and a break in the continuity permits the escape of lines of magnetic force into the surrounding medium.

209. Magnetic induction. We have already spoken of magnetisation produced by induction or induced magnetisation: the term magnetic induction is however used with a special significance.

If a piece of iron is placed in a magnetic field a number of lines of force are concentrated into the iron and after traversing it pass out again into the air.

At points where the lines of force enter the iron we have a series of south poles, at points when they leave we have north poles, in each case formed by induction.

The force both throughout the iron and in the surrounding air depends in part on this induced magnetism, and this again depends on the strength of the original field and on the shape and magnetic quality of the iron.

Again, if we suppose a cavity formed in the iron, Fig. 215, the force on a unit north magnetic pole placed in the cavity will depend in part upon the original strength of the field, in part on the shape of the cavity.

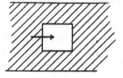

Fig. 215.

Now let H be the strength of the field and let us consider a portion of the surface of the iron where the lines of force of the field enter it at right angles. South polar or negative magnetism will be induced on this area, and the amount of this will depend on H and on the iron. Let us suppose the quantity of south polar magnetism induced per unit of area to be I and let us put $I = \kappa H$, then κ is called the coefficient of induced magnetisation. Experiment shews that κ is not independent of the strength of the magnetising field; it depends on this and on the past history of the iron.

210. Magnetic force within a mass of iron.

Now let us take a case in which the field is uniform in the iron and imagine a long narrow cylindrical cavity as in Fig. 216, cut in the iron with its axis in the direction of the magnetising force and its ends at right angles to that direction.

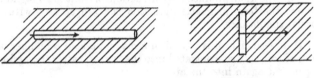

Fig. 216.   Fig. 217.

On one end of this cavity where the lines of force leave it we shall have an amount of induced magnetism I or κH per unit of area, on the other end there will be $-\kappa H$ per unit of area. Thus if the ends be a square centimetres in area the quantities of magnetism on them will be Ia and $-Ia$ respectively. Since the curved walls of the cavity are parallel to the lines of force of the field no lines of force either enter or leave these, thus no induced magnetism is on them. Thus the force within the cavity is the original magnetising force H together with the forces arising from the quantities of magnetism Ia and $-Ia$ on the ends. Now if the cavity be very narrow a is very small, and the quantities of magnetism concerned will be very small, while if the cavity be very long they will be at a great distance from the point at which the force is being estimated, thus the amount they contribute to the force will be vanishingly small. Hence the resultant magnetic force at a point within the cavity will be H, the magnetising force.

211. Magnetic force in a crevasse. Now let us

suppose that the cavity is extremely short, a kind of narrow crevasse, cut across the lines of force as in Fig. 217. Then the two ends with the magnetic charges aI and $-aI$ will be like two small oppositely charged flat plates close together charged to surface densities I and $-I$ respectively. They resemble the plates of an electric condenser.

Now we have already seen, Section 43, that if σ be the surface density of a condenser with air for its dielectric the force between the plates is $4\pi\sigma$, and that it acts from the positive to the negative plate. But magnetic forces follow the inverse square law in the same way as electrical. Hence we can infer by exactly similar reasoning that the force in a crevasse cavity cut normal to the magnetising force due to magnetism with density I and $-I$ on its flat faces is $4\pi I$ acting from the positive to the negative face.

In order to get the actual force in the crevasse we must add to this force $4\pi I$ the magnetising force H; and we have as the result $H + 4\pi I$, or putting for I its value κH we get as the force $H + 4\pi\kappa H$ or $H(1 + 4\pi\kappa)$.

This quantity, the resultant magnetic force in the crevasse cavity, is known as the **Magnetic Induction**, and is usually denoted by B. Thus if we write $1 + 4\pi\kappa = \mu$ we have

$$B = (1 + 4\pi\kappa) H = \mu H.$$

The quantity μ is known as the permeability.

We are thus led to the following definitions.

DEFINITION. *The ratio of the quantity of magnetism induced per unit of area on a surface normal to the direction of the magnetising force to that force is known as the* **Coefficient of induced magnetisation.**

It is denoted by κ, and we have $I = \kappa H$.

DEFINITION. *The resultant magnetic force within a narrow crevasse cut in a magnetic medium and bounded by faces perpendicular to the magnetising force is known as the* **Magnetic Induction.**

It is usually denoted by B, and we have

$$B = H + 4\pi I = H(1 + 4\pi\kappa).$$

DEFINITION. *The ratio of the magnetic induction to the magnetising force is known as the* **Permeability.**

It is usually denoted by μ.

Hence $\qquad\qquad B = \mu H.$

212. Magnetic permeability. We may obtain these results somewhat differently thus.

Consider a bar magnet, uniformly magnetised, let I be the intensity of its magnetisation, a the area of its cross-section, then the strength of its north pole is Ia. Now from the definition it can be shewn that from a pole of strength m the number of lines of force which issue is $4\pi m$. Again, the magnetic force at any point is the number of lines of force which cross unit area placed at right angles to the force in such a position as to contain the point.

Now imagine the magnet bent round into the form of a ring so that its north and south poles may almost coincide. It will produce no external magnetic field except just in the gap between the poles, and the number of lines of force which traverse the gap is $4\pi Ia$. Hence the number of lines of force per unit area in the gap is $4\pi I$, and this is the magnetic force in the gap.

If the magnetisation of the magnet be produced by induction we must add to this force H the magnetising force, and we get as the resultant force in the gap $H + 4\pi I$.

If this force be defined as the induction and denoted by B then

$$B = H + 4\pi I.$$

But $\qquad\qquad I = \kappa H.$

Hence $\qquad\qquad B = H + 4\pi\kappa H$

$$= (1 + 4\pi\kappa)\, H$$

$$= \mu H.$$

Thus the magnetic induction in any medium is the number of lines of force which cross per unit area a narrow gap in the medium, the surfaces of the gap being at right angles to the lines of force.

It is clear from the above that μ is unity and the measure of the induction the same as that of the force for all media for which κ vanishes, that is to say for all except magnetic media. Experiment shews that κ is excessively small, practically zero, except in the case of iron, nickel, and cobalt. The value of μ depends on the magnetising force. For iron in moderate fields it may range between 400 and 2500, for nickel and for cobalt its maximum limit is about 200.

In certain media κ is a small negative quantity; when this is the case μ is less than unity; such media are called diamagnetic. Iron and media for which κ is positive and μ greater than unity are called paramagnetic.

In diamagnetic media the induced magnetism is opposite to the magnetic force.

213. The magnetic circuit. Now we know from Ohm's law that if C be the current of electricity which crosses unit area of a conductor whose conductivity is k and in which the electric force is E, then

$$C = kE.$$

We may compare this with the equation

$$B = \mu H,$$

connecting together the induction or flow of magnetic lines of force per unit area, the permeability and the magnetising force, the two are clearly analogous. If we treat the magnetic induction as a flux or current per unit area it is related to the permeability, and the force, in the same way as the current per unit area is related to the conductivity and the electric force. We may speak of the magnetic circuit in the same way as we use the expression the electric circuit. There is, however, this great difference that k, the conductivity, does not depend on E, the electric force, while μ, the magnetic permeability, does depend on H. Still the analogy is often useful.

214. Magnetic Reluctance. Again, let us consider a magnetic circuit of uniform section a and let it be subject to a uniform force H. Let l be the length of the circuit and

Ω_1, Ω_2 the magnetic potentials at its two ends, then since the force is the space-rate of change of potential, we have

$$H = (\Omega_1 - \Omega_2)/l.$$

Let $\bar{B}$ be the total flow of induction through the circuit.

Then $$\bar{B} = Ba.$$

Hence total flow of induction

$$= \bar{B} = Ba = a\mu H$$
$$= \frac{\mu a}{l}(\Omega_1 - \Omega_2).$$

But if C be the total flow of electricity in a circuit of section a, conductivity k, and length l, V_1, V_2 being the electric potentials at the ends,

then $$C = \frac{ka}{l}(V_1 - V_2),$$

while l/ka is the resistance of the conductor.

Thus if we define $l/\mu a$ as the magnetic resistance or reluctance of the circuit, we can apply the laws governing the flow of electricity to magnetism. For example, if we suppose a body is composed of a series of portions of lengths l_1, l_2, ... sections a_1, a_2, ... and permeabilities μ_1, μ_2, ..., then the total reluctance is

$$l_1/\mu_1 a_1 + l_2/\mu_2 a_2 + \ldots$$

and the total flow of induction, if Ω_1, Ω_2 are the potentials at the beginning and end of the circuit, is

$$(\Omega_1 - \Omega_2)\bigg/\bigg\{\frac{l_1}{\mu_1 a_1} + \frac{l_2}{\mu_2 a_2} + \ldots\bigg\}.$$

Thus it is clear that if we wish to produce, for a given difference of magnetic potential, a large flow of induction we must keep the reluctance small.

Now we have already stated that for a given magnetising force the permeability of iron is enormously greater than that of any other medium, so that to secure large inductions we must use iron and keep the air gaps in our circuit as narrow as possible, that is, keep values of l which correspond to the small values of μ very small.

Thus we see how it is that by putting an iron core into a spiral carrying a current we increase enormously the magnetic force at the ends of the spiral, the magnetic reluctance of the circuit is reduced and the flow of induction increased.

215. Measurement of magnetic permeability.
Several methods have been devised for the measurement of κ and μ. Some of these can be best described after we have considered the phenomena of electromagnetic induction. In the magnetometric method the arrangements described in Section 92 for the determination of M/H are employed.

The iron to be magnetised takes the form of a thin rod and is placed inside a magnetising spiral. When a current passes it becomes a magnet and will deflect a small magnet placed near. From this deflexion the magnetic moment induced by the current can be found in terms of the inducing force; but this force is known if the current be known and hence κ and μ can be determined.

The experiment usually is conducted thus.

EXPERIMENT 53. *To determine by means of the magnetometric method the coefficient of induced magnetisation, or susceptibility, and the permeability of an iron rod.*

The iron rod NS, Fig. 218, which should be about 40 cm. long and 1 to 2 millimetres in diameter is placed inside a magnetising spiral consisting of a thin coil some 50 centimetres in length wound with one or two layers of insulated wire. The iron rod lies along the axis of the coil and this is directed east and west.

A small magnetometer, usually one with a mirror magnet, is placed at a point on the axis produced distant r centimetres from the centre of the rod.

On passing a current through the coil the iron rod becomes a magnet and the magnetometer needle is deflected. Let ϕ be the angle of deflexion, M the magnetic moment of the induced magnetisation, $2l$ the length, and A the diameter of the rod, which we suppose to be circular in section.

Then if F be the strength of the field at the magnetometer due to the magnetised iron, H the strength of the earth's field, we know that, since the directions of F and H are at right-angles,

$$F = H \tan \phi.$$

Again, the strength of each pole of the iron rod is $M/2l$, one pole is at a distance $r - l$, the other at a distance $r + l$ from the magnetometer, so that if we suppose F is due entirely to the magnetism of the rod, we have

$$F = \frac{M}{2l} \left\{ \frac{1}{(r-l)^2} - \frac{1}{(r+l)^2} \right\} = \frac{2Mr}{(r^2 - l^2)^2}.$$

Hence we have

$$M = \frac{(r^2 - l^2)^2}{2r} H \tan \phi.$$

But M is due to the magnetism induced in the rod by the current in the spiral. Let X be the magnetic force at any point in the axis of the spiral due to this current and let I be the induced magnetisation. We assume this force to be the same along that part of the axis which is occupied by the rod. Then, if κ is the coefficient of induced magnetisation or the susceptibility, we have

$$I = \kappa X.$$

Since the area of the end of the rod is πa^2 the strength of each pole is $\pi a^2 I$ and M the magnetic moment of the rod is therefore $2\pi a^2 l I$ or $2\pi a^2 l \kappa X$.

Substituting this value for M we have

$$\kappa = \frac{(r^2 - l^2)^2}{4\pi r a^2 l} \frac{H}{X} \tan \phi,$$

but if N be the number of turns per unit length of the spiral then since X is due to the current i amperes we have (§ 201) $X = 4\pi Ni/10$.

Now i can be measured by an ammeter or in some other convenient way, thus X is known and substituting for it we have finally

$$\kappa = \frac{10 (r^2 - l^2)^2}{16\pi^2 a^2 r l} \frac{H}{Ni} \tan \phi.$$

One or two points require to be noticed.

In the first place the field F is not due solely to the action of the iron rod; the current in the spiral would, even if the rod were removed, produce an effect on the magnetometer. This can be measured or allowed for by making an experiment without the rod in position; the deflexion observed will then be due to the current only; it is better however to compensate for it by inserting in the current circuit a few turns of wire near the magnetometer. These can easily be adjusted so that the effect they produce on the magnet shall be equal and opposite to that of the spiral. To do this the iron is removed and the position of the additional turns altered until no effect is observed on the magnetometer whatever be the current in the circuit. Then when the iron is in position the whole effect is due to its induced magnetisation. Again, it has been assumed that the magnetising field is solely that due to the current; but when the iron becomes magnetised the field in its interior is in part due to its own magnetisation and thus part of the field is opposite in direction to the magnetising field, for if the latter act from S to N (Fig. 218) then S becomes a south pole, N a north, and the field in the magnet due to these poles is from N to S.

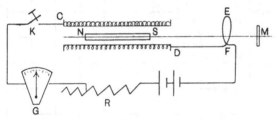

Fig. 218.

Thus the magnetising field is not strictly that due to the current, as has been assumed. The difficulty is met to some extent by making the rod long and thin.

The arrangement of apparatus is shewn in Fig. 218, in which NS is the rod, CD the magnetising spiral, EF the compensating coil, M the magnetometer, R is a resistance for regulating the current, G the ammeter, and K a key, preferably a reversing switch, for making and breaking the circuit.

216. Curve of magnetic induction. When the value of κ has been found thus, the value of μ can be calculated by the formula $\mu = 1 + 4\pi\kappa$ and B is given by multiplying μ by the magnetising force X.

A series of experiments can be made by gradually reducing

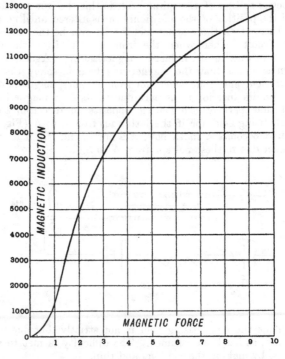

Fig. 219.

the resistance in circuit and thus increasing the magnetising force; if this be done the results may be plotted as a curve in which the ordinates represent B and the abscissae the values of X, the magnetising force; if the iron be initially unmagnetised, and this must be tested for in the usual way, the curve will have the form given in Fig. 219.

If at any stage the current be broken the iron will remain magnetised, but the amount of this permanent magnetisation will depend largely on the method adopted for breaking the current, still the residual magnetic moment

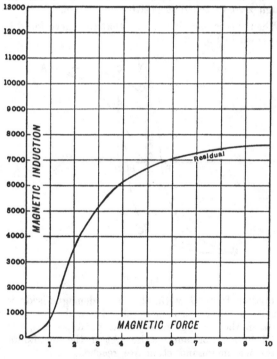

Fig. 220.

and the residual induction can be calculated, and if this be done for various values of X a curve of residual magnetisation can be found as in Fig. 220.

217. Hysteresis. The behaviour of iron in a magnetic field can be more completely investigated by carrying the current through a complete cycle. Starting from zero current the resistance in the circuit is gradually diminished, thus increasing the current until the maximum value desired is reached. The current is then reduced to zero, after which its direction is reversed and it is carried on by similar stages until the former maximum—only with the direction of flow changed—is reached. From this point it is carried back through zero to the first positive maximum.

When this is done it is found that the curve for B has the form shewn in Fig. 221.

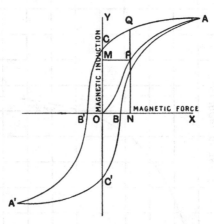

Fig. 221.

Starting from O with the iron demagnetised, it rises gradually at first, then more steeply, and after becoming very steep the slope gradually falls until the line is almost parallel to the horizontal axis. At A the maximum force and the maximum induction are reached.

The force is now reduced and the induction falls, but it does not return along the line APO but along a line ACB', lying above it and to the left, cutting the vertical axis in C and the horizontal in B'.

The induction for each value of the force is greater than the value it had for the same force on the outward journey. Thus let QPN parallel to OY meet OX in N, then when the iron was first being magnetised, for the force ON the induction was PN; when the magnetisation is being reduced from that corresponding to A, for the same force ON the induction is QN, which is greater than PN and corresponds on the outward curve to a force greater than ON.

The induction lags behind the magnetising force; when the force has reached zero the induction is CO and the force has to be made negative and equal to OB' before the induction becomes zero. As the force increases still further negatively the induction becomes negative until when the negative maximum is reached for the force the point A' of the diagram represents the condition of the iron. As the force again returns to zero and after passing it approaches its former maximum at A, the curve $A'C'BA$ is traced by the induction and this curve is found to be symmetrical with respect to the axes with the position $ACB'A'$.

The name hysteresis—a lagging behind—has been given to this phenomenon because in all cases the induction for a given value of the force lags behind the value it would have for that same value of the force if the original curve of magnetisation from the demagnetised condition had been followed.

We might have drawn a similar curve for the induced magnetisation and drawn similar conclusions from its form.

218.   Theories of Magnetisation.   Ewing's Model, Section 88, enables us to understand some of the changes which probably go on in a piece of iron which is being magnetised, and indeed, it has been shewn by direct experiment, that an assemblage of a large number of small compass needles behave, when subject to a gradually increasing magnetic field, exactly in the same way as the bar or rod of iron.

At first the magnets are arranged in closed circuits; the assemblage produces no external field. When a small magnetic force is applied some few of the needles are disturbed, and on the whole, there is a tendency for the needles to set themselves parallel to the field; but the mutual forces between the needles restrain this tendency, and the magnetic moment of the bar remains small. As the force is increased, it gradually overcomes the mutual attractions; more and more of the circuits are broken up, and for certain values of the external field depending on the strength of the mutual action between the compass needles, this destructive action goes on very rapidly, in other words the induction increases very rapidly with the force, we are on the steep part of the curve. For still larger values of the force nearly all the circuits have been broken, the increase of induction becomes less rapid until a condition is reached when there are no groups left, the axes of all the magnets are parallel to the field, the iron is magnetised to saturation.

Again, it is clear that the configuration of the magnets, corresponding to any given value of the force, depends in part upon the force, in part upon the configuration when the force was applied.

If we imagine the assemblage demagnetised when the force is applied, we shall get one result depending on the balance between the impressed force and the mutual forces between the magnets; if we imagine the assemblage saturated and the external field to be then reduced until it has the same value as in the previous case, the configuration will usually be different, because the mutual forces which have been overcome are different.

The configuration of the magnets at any moment depends upon the external field, and upon the configuration when the field was applied. In this way the model enables us to explain the phenomenon of hysteresis.

219. Energy needed to Magnetise Iron. Work must be done to magnetise a piece of iron, and this for two reasons at least, (1) a magnetic field is produced and the magnetised iron can do work, and (2) energy is necessary to break up the closed molecular circuits in the iron, to move the molecules against their mutual attractions.

By the demagnetisation of the iron the first part of the energy can be recovered. Not so however with the latter portion. As the molecular chains break up the molecules are thrown into a state of disturbance, oscillating backwards and forwards, and part of the energy supplied is frittered away as heat. This part of the energy is irrecoverable as magnetic energy, the process is irreversible; by carrying the iron through a complete cycle an amount of heat is produced which it can be shewn is measured per unit of volume by the area of the hysteresis loop divided by 4π.

We can obtain a connexion between the area of the loop and the work done in magnetising the iron, thus:

Consider a small magnet, let m and $-m$ be the strengths of its two poles, H the field in its neighbourhood, and let us consider the work required to increase the pole strengths by small amounts m' and $-m'$ respectively. Work will be done in carrying the amount $-m'$ to the south pole from outside the field, and also in carrying $+m'$ to the north pole. As we have seen, when discussing the theory of the potential, the work will not depend on the path followed. Now we may imagine m' brought to S by exactly the same path as was travelled by $-m'$ and then carried directly from S to N. The amounts of work done in carrying m' and $-m'$ to S will be equal and opposite, the amount of work done in carrying m' from S to N is $2m'Hl$, where $2l$ is the length of the magnet SN.

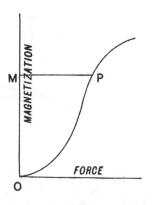

Now the poles are increased in strength by increasing the induced magnetisation I and if this be increased by a small amount δI and if a be the area of a section of the magnet normal to the axis, then $m' = a\delta I$.

Hence the work done

$$= 2la\ H\delta I.$$

But $2la$ is the volume of the magnet, hence the work done per unit of volume is $H\delta I$.

Hence looking upon the final state as having been reached by successive additions to the magnetisation, we see that the work

Fig. 222.

done may be written $\Sigma H\delta I$ where Σ is an abbreviation for the sum of quantities such as the above.

But if we draw a curve OP, Fig. 222, such that the abscissa of a point P represents the force H and the ordinate the magnetisation I, then as in Section 37 we can shew that the area OPM, where PM is parallel to the line of force, is equal to $H\delta I$. Hence the area OPM measures the energy per unit volume required to magnetise the iron.

Now in Fig. 223 let $A'C'BA$ be the curve of magnetisation starting from the point A' at which the force and magnetisation have their maximum negative values, $ACB'A'$ the curve of demagnetisation and let AM, $A'M'$ be parallel to OX.

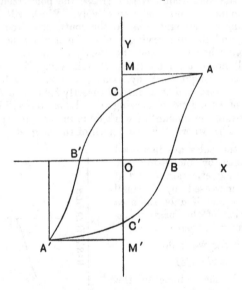

Fig. 223.

Then in passing from C' to A an amount of energy $C'AM$ is absorbed per unit volume, but in passing from A to C an amount CAM is returned. Thus an amount CAC' is absorbed in passing by the curve from C' to A and then to C. Similarly the amount absorbed in passing from C to A' and then to C' is $C'A'C$. Hence the amount absorbed in going round the

cycle is represented by the area of the loop. It should be
noted that in this curve the ordinates represent the induced
magnetisation I, not the induction B as in Fig. 221, and the
energy absorbed is equal to the area of the magnetisation loop.

We obtain the induction curve from this one thus :

Since we have $B = H + 4\pi I$,

we can draw the magnetisation curve.

Let $P'PN$, Fig. 224, be parallel to OY and let $Q'Q$ be the
points on the induction curve corresponding to P' and P.

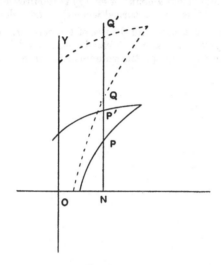

Fig. 224.

Then since $B = H + 4\pi I$

we have $QN = ON + 4\pi PN$

$Q'N = ON + 4\pi P'N$.

Hence $QQ' = 4\pi PP'$.

Thus if we draw consecutive ordinates it is clear that the
areas of corresponding strips of the curves are as 1 to 4π, for
their breadths are equal and their heights are as QQ' to PP'.

Thus the area of the induction loop is 4π times that of the magnetisation loop.

Hence the area of the hysteresis curve as defined in Section 217 is 4π times the energy per unit volume required to carry the iron round the cycle.

When the area of the loop is small we infer that the molecules follow the changes of the magnetising force readily; all the energy spent in magnetisation can be obtained by demagnetising the iron. When on the other hand the loop is large, an additional amount of energy is required to magnetise the iron and this is not returned when the iron is demagnetised.

The iron is heated and the rise of temperature is a measure of this hysteresis loss, which can be calculated by measuring the area of the loop and the volume of the iron.

CHAPTER XXII.

ELECTROMAGNETIC INSTRUMENTS.

220. Moving Coil Galvanometers. The mutual action between a wire carrying a current and a magnetic field, or between a similar wire and a piece of soft iron, is made use of in many forms of instrument for the measurement of current or of electromotive force.

We have already seen how to utilize it in the ordinary galvanometer in which the current circulates in a fixed coil and a magnet is delicately suspended near its centre. In some more modern forms of instruments—moving coil galvanometers—the magnet is fixed, and the coil carrying the current moves.

Such an instrument is shewn in Fig. 225. The magnet is in the form of a powerful horseshoe magnet, the poles of which are brought near together. The coil of wire hangs in the field between the poles, its plane being parallel to the lines of force, and the axis therefore of the magnet, to which it is equivalent when a current is passed through it, is at right angles to the lines of force. When a current circulates in the coil it tends to turn so as to include in its area the maximum number of lines of force; if the coil were suspended quite freely it would turn through a right angle, and set itself perpendicularly to its former position. But the wires suspending the coil and by means of which the current is brought to it, exercise a constraint on the coil, and in consequence it only turns until the couple due to the electromagnetic action balances that due to the constraint. In some instruments the coil hangs by a single wire, a second wire is continued downwards below the coil, being stretched fairly tight. The coil in turning twists this wire and the constraining couple arises from the torsion thus produced. The current

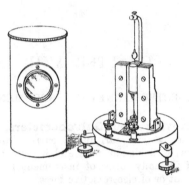

Fig. 225.

passes in from above and out below the coil. In very sensitive instruments the lower wire is loose, the couple arises from the torsion of the upper wire which is stretched by the weight of the coil.

In other instruments again the suspension is bifilar; the coil hangs from two very fine parallel wires, the current enters by one and leaves by the other; as the coil turns it is lifted slightly by the action of the wires and its weight supplies the restraining couple.

The motion of the coil may be indicated by a pointer moving over a scale, or by a ray of light reflected from a mirror attached to the coil.

The relation between the strength of the current and the deflexion will depend on the distribution of the lines of magnetic force in the field. By properly choosing the form of the magnet this can be arranged so that the deflexion is very accurately proportional to the current; in such an instrument the current causing a given deflexion is read off directly.

In another form of the arrangement, the coil is mounted on pivots between two jewels, and carries a pointer which moves over a uniformly divided scale, divided so as to read amperes directly. Such an instrument is shewn in Fig. 226.

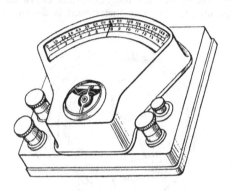

Fig. 226.

221. Soft iron Ammeters. In another class of instruments the attraction between a coil carrying a current and a piece of soft iron is used to measure the current. A piece of soft iron is arranged so that it can be sucked up into the core of the coil when the current passes; the distance the iron moves is indicated by the motion of a pointer attached to it. The pointer moves over a scale and the current in amperes is read directly. The iron is prevented by its weight from being sucked completely into the coil and the equilibrium position of the pointer is reached when the couple due to the attraction between the coil and the iron balances that due to the weight of the iron.

If the susceptibility of iron were constant, so that the induced magnetisation was always proportional to the current, then since the force between the iron and the coil in any given position depends on the product of the strength of the current and the induced magnetisation, and this latter is by hypothesis proportional to the current, the force will depend on the square of the current; from this fact the instrument could be graduated, but since the susceptibility is not constant, but itself depends on the current, such an instrument has usually to be graduated by direct experiment, employing calculations based on the magnetisation curve as the basis of the experiment. Such an instrument is shewn in Fig. 227.

Fig. 227.

222. Ammeters and Shunts. When an ammeter is used for the measurement of a large current only a small fraction of the current passes through the coils of the instrument. A shunt of small resistance is employed to connect together the poles of the instrument, and the main portion of the current flows through the shunt. A definite fraction of the current depending on the ratio of the resistance of the shunt to that of the coil, passes through the coil, and by measuring this the whole current is estimated.

For instance, the instrument may be arranged to have a resistance of 1 ohm, and be such that when a current of 0·1 ampere passes through the coils the deflexion is 100 scale divisions. To produce a current of 0·1 ampere a potential difference of 0·1 volt is required in this case, and each division of the scale registers the passage of a milliampere or the application of a millivolt.

Now let us shunt the poles A, B of the ammeter, Fig. 228, by 1/99 of an ohm; then if a current of ·1 ampere enter at A one hundredth of the current will pass through the coil C and 99/100 through the shunt. This follows from Section 159,

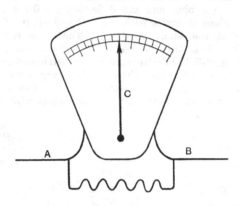

Fig. 228.

for we have seen that if S be the resistance of the shunt, G that of the galvanometer, then the current through the galvanometer coil is $S/(S + G)$ of the whole current.

Hence in the case supposed the sensitiveness has been reduced one hundredfold; the deflexion for a total current of 0·1 ampere would only be one division, and the instrument would read up to $100 \times ·1$ amperes or 10 amperes. By decreasing the resistance of the shunt we increase the range. For measuring high currents we should probably start with a less sensitive instrument, one in which for example an E.M.F. of a volt produced a deflexion of 100 divisions, then if the shunt were ·001 ohms in resistance, and a current of 1000 amperes traversed it, the E.M.F. between the poles would be 1 volt and the deflexion 100. The total current would be rather greater than 1000 amperes, because of the small fraction of the whole which traverses the coil of the ammeter, but by using a shunt of slightly greater value, the instrument could be adjusted so that 1000 amperes corresponded to 100 divisions;

thus its range, assuming one division to be legible, would be
from 10 to 1000 amperes.

223. Voltmeters. The same instrument can be used
as a voltmeter. For let us take the first case when the coil
resistance is 1 ohm and the deflection for 0·1 volt is 100.
Place 99 ohms in series with the coil and apply 10 volts to
the ends of the whole resistance. The E.M.F. between the
poles will still be 0·1 volt, the current 0·1 ampere, thus the
deflexion is still 100 divisions, and the instrument reads from
0·1 volt up to 10 volts. Similarly by using a less sensitive
instrument and higher resistances we can measure higher
voltages. Fig. 229 shews such an instrument with its shunts

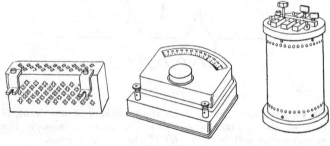

Fig. 229.

and resistances. The shunts, one of which is shewn to the left,
are strips of manganin or some similar resistance material
arranged so as to have considerable surface and therefore to
heat but little with the passage of a considerable current; the
volt resistances on the right hand of the figure are coils wound
as in an ordinary resistance box.

It should be noticed that the readings of such an instrument
are to some extent affected by temperature. They depend on
the ratio of two resistances; one of these, the shunt or the
volt resistance, is usually of manganin or some material which
does not change much in resistance with temperature, the
other, the coil, is of copper and has a considerable temperature
coefficient.

The range and applicability of the soft iron instruments can be extended by the use of shunts or volt resistances in the same manner.

224. The electrodynamometer. In this instrument, which may take various forms, one of which is shewn in Fig. 230, two coils, one fixed, the other moveable, are used.

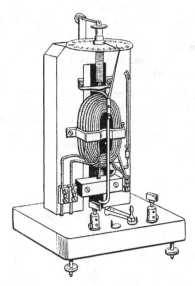

Fig. 230.

The centres of the two coils coincide and their axes are at right angles so that if currents circulate in the two, their lines of force at the common centre are perpendicular, and the moving coil tends to move so that its axis and its field may be parallel to those of the fixed coil respectively. The moving coil is suspended in such a way as to resist this force, either by means of a wire having torsion or by a bifilar suspension, and if the instrument be once graduated, then by measuring the angle through which the coil is deflected the current can be calculated.

In practice, instead of allowing the coil to be deflected the suspension head is turned in the opposite direction to that of the deflexion, and the coil is thus brought back to its equilibrium position. The angle through which the suspension head is turned measures the couple required to hold the coil in its equilibrium position and enables the current to be calculated.

For let i_1, i_2 be the currents in the two coils; the strength of the field due to the fixed coil is proportional to i_1, the magnetic moment of the moving coil to i_2, and the couple on a magnet of moment M in a field of strength H is MH. Hence the couple on the moving coil is proportional to $i_1 \times i_2$. Thus the deflexions are proportional to $i_1 \times i_2$.

If then i_1 be a known constant current the deflexions will be proportional to i_2 and the current can be measured. More usually however the same current i circulates in the two coils, thus i_1 and i_2 are equal, and the deflexions are proportional to i^2. Hence i the current is measured by the square root of the deflexion.

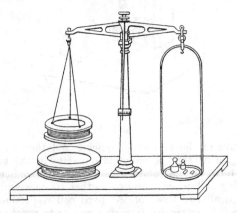

Fig. 231.

225. The Ampere balance. This instrument in its simple form, shewn diagrammatically in Fig. 231, has a coil with its plane horizontal suspended from one end of the beam of a balance. A fixed coil with its plane also horizontal rests below this and the weight of the hanging coil is counterpoised by weights in the pan. Flexible connexions are arranged so that a current can be passed through the moveable coil ; if the same current be passed through the fixed coil also there is an attraction—or it may be a repulsion—between the two which is proportional to the product of the moments of the two equivalent magnets. As each of these is proportional to the current the force of attraction is proportional to the square of the current. To restore equilibrium weights must be placed in the pan of the balance, and the current in any case will be proportional to the square root of the weight. Thus two currents can be compared by comparing the square roots of the weights used in the two cases, or if the weight required to balance some given current, one ampere say, be known and be equal to W_0, where that required for a current i is found to be W, then i is given by the equation

$$i = \sqrt{\frac{W}{W_0}} \text{ amperes.}$$

The standard ampere balance of the Board of Trade is constructed in this manner ; the weight required to produce equilibrium when a current of one ampere is circulating in the coils was determined with great care and so long as the construction of the balance does not change in any way this weight remains fixed.

The ampere is now defined for legal purposes as the current which must be passed through the coils of the instrument to balance this weight.

226. Lord Kelvin's Ampere balance. The most usual form of current balance, however, is that of Lord Kelvin.

One of these is shewn in Figs. 232 (a) and (b) of which Fig. 232 (a) gives the connexions and indicates the path of the current while 232 (b) is a drawing of the complete instrument. Two moveable coils are attached, one to each end of a balance beam ; both beneath and above these are fixed coils, and the direction of the current is such that the electromagnetic force on one of the moveable coils is upwards while on the other it is downwards.

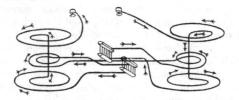

Fig. 232 *a*.

A moveable weight can slide on a graduated horizontal arm attached to the beam, and a pointer at the end of this arm shews when the beam is in its horizontal position.

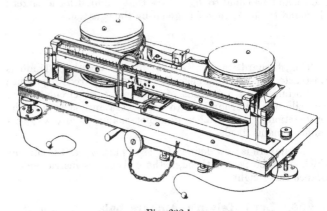

Fig. 232 *b*.

On passing the current through the coils the beam is deflected, and the weight must be shifted to bring the pointer back to its sighted position.

The couple deflecting the beam is proportional to the square of the current, that restoring it is proportional to the distance the weight has been shifted from the fulcrum.

Hence for a given weight, the current is proportional to the square root of the distance between the weight and the fulcrum, and by graduating the arm along which the weight slides according to a square root scale, the currents can be read directly from the position of the weight.

The current is introduced into the instrument by flexible leads, which do not interfere with the action. In Lord Kelvin's instruments, these form the ligaments by which the beam is suspended. The value of the weight is of course determined by an experiment with a known current; by employing a series of weights the range of the instrument is extended.

227. Measurement of Alternating Current. It should be noticed that the indications of these last three instruments depend on the square of the current. Thus by reversing the current, so long as its magnitude is unaltered, the indication of the instrument is unaffected. In many applications of electricity alternating currents (see Section 246) in which the direction of the current is being continually reversed are employed. These instruments may with proper precautions be used to measure such currents.

228. Influence of external fields. The indications of any of the above instruments will be affected to some extent by the strength of the external magnetic field in which they happen to be placed; in the case of a moving coil galvanometer, for example, the deflexion depends on the strength of the field in which the coil hangs, and so with the others; but since the field of the permanent magnets is very intense compared with the earth's field, for example—(in a good instrument the field may be 800 to 1000 units, while the earth's field is ·18)—the effect of the earth's field is very small, and that of small changes of that field is smaller still. While small changes in the direction of the earth's horizontal force produce a considerable effect on the zero of an ordinary astatic galvanometer, they are practically inappreciable with a suspended coil instrument. Indeed such an instrument can be employed in the neighbourhood of electromagnetic machines, such as dynamos or motors (see Section 240) without its indications being very seriously affected, in positions in which an ordinary moving magnet galvanometer would be useless. The reason is clear; in the suspended coil instrument the deflexion is increased by increasing the field strength of the instrument. This therefore is made so strong that external changes are masked. In a moving magnet instrument the sensitiveness is increased by weakening the field of the instrument, small changes, then, in the weak field due to external influences produce large effects.

CHAPTER XXIII.

ELECTROMAGNETIC INDUCTION.

229. Faraday's Experiments on Induction. We have already seen, Section 199, that when a circuit carrying a current is placed in a magnetic field a force acts on it and tends to move it into such a position that a maximum number of lines of induction due to the field may pass through the circuit from its south face towards its north face.

We proceed now to deal with some phenomena, first observed by Faraday, which may be looked upon as the converse of the above.

Faraday shewed that whenever a circuit in a magnetic field was moved relatively to the field so as to vary the number of lines of magnetic induction which traverse it, then an electromotive force is set up round the circuit causing a current to flow which lasts so long as the number of lines of induction linked with the circuit is varied. The relative motion of the field and the circuit may be produced by moving either the circuit or the magnetic system to which the field is due.

Moreover it should be noted that the currents set up are always such as to oppose by their electromagnetic action on the field the motion to which they are due.

These currents are said to be induced.

EXPERIMENT 54. *To investigate the production of an induced current by the motion of a magnet.*

The ends of a coil of wire, Fig. 233, are connected to a galvanometer which is placed at some distance from the coil.

A galvanometer may if necessary be constructed for the purpose by utilizing a similar coil. This is placed in the magnetic meridian and a compass needle is pivoted at its centre; when undisturbed the compass-needle rests with its axis in the plane of the coil, on passing a current through the coil the needle is deflected.

Fig. 233.

Take a long bar magnet and move its north pole up to the coil of wire, passing it through the coil, the galvanometer needle is deflected shewing a current through the circuit, as soon however as the motion of the magnet ceases the needle comes back to its zero position, thus the current only lasts while the motion continues.

Withdraw the magnet from the coil. The galvanometer needle is again deflected, but in the opposite direction.

Repeat the experiment, but now bring up the south pole of the magnet instead of the north, the galvanometer needle is deflected as in the last case; the approach of the south pole therefore causes a current in the same direction as the withdrawal of the north pole and *vice versâ*.

The direction of the current can be inferred from the direction of motion of the magnet. To make things clear suppose the coil placed with its plane vertical and north and south and connected with the galvanometer coil in such a way that the current may circulate in the same direction round the two, and let the north pole of the magnet approach the coil from the east side.

Then it will be found that the north pole of the galvanometer needle is deflected to the east. From this we infer that the current in both coils passes from south to north below the magnet, Fig. 234. Thus if the plane of the paper represent the face of the coil of wire which the magnet

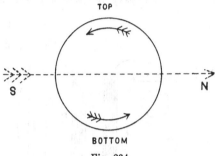

Fig. 234.

approaches, so that the upper side of the paper is east, the right hand being north, the current is as indicated in the figure. Thus the upper surface of the paper becomes the north face of the magnet which is equivalent to the circuit, and the motion of the magnet inducing the current is opposed by the repulsion between the circuit and itself.

On withdrawing the north pole it will be found that the direction of the current is reversed, the upper face of the circuit is a south face and the north pole is attracted by this, the electromagnetic forces again oppose the motion.

It is clear from the principle of the conservation of energy that this must be so; for the current possesses energy and this energy it obtains from the work done in moving the magnet towards the circuit in opposition to the electromagnetic repulsion, or withdrawing it from the circuit in opposition to the electromagnetic attraction.

Corresponding effects are produced if the magnet be kept fixed and the coil moved up to it.

The above effects have been produced by the motion of a magnet. For the magnet we may substitute a coil of wire carrying a current.

EXPERIMENT 55. *To investigate the production of an induced current by the change in the strength of a neighbouring current.*

The same apparatus is required as in the preceding experiment, except that for the magnet we substitute another coil of wire I. connected to a battery and to a key, Fig. 235. This

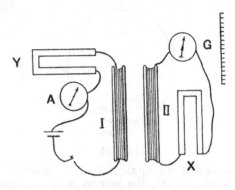

Fig. 235.

second coil is placed near to the first, the planes of the two being parallel. As previously, the coils I. and II. should be placed at some distance from the galvanometer G so as to avoid any direct magnetic action on its needle. The coil connected to the battery is spoken of as the primary coil, that connected to the galvanometer as the secondary coil.

Resistance boxes X and Y are conveniently included in the circuits in order to vary the currents if desired. The primary circuit also conveniently includes an ammeter A.

Make contact in the primary circuit and allow a current to pass; the galvanometer needle is temporarily deflected, but if the key be held down it returns at once to its zero position. Break contact at the key; the needle is again temporarily deflected, but in the opposite direction to its previous motion.

The first contact has produced a magnetic field in the neighbourhood of the primary coil. Lines of magnetic induction well out from the coil, some of these thread the secondary, thus the number of lines of induction linked with the secondary is varied and an induced current is the consequence.

If the distance between the two coils be increased the effect is diminished, if on the other hand a bundle of iron rods is inserted in the secondary, the effect is to increase greatly the induction through it, and the induced currents are larger. Non-magnetic materials placed in the coils do not change the result.

In some cases we can calculate the total change of induction through the secondary; if the galvanometer be a suitable one arranged for ballistic work, we have seen already, when treating of condensers, that the first throw of the needle is proportional to the total quantity of electricity which circulates round the coil, and it is not difficult to arrange experiments to prove that this total quantity is proportional to the total change of induction and inversely proportional to the total resistance of the circuit[1].

Thus the total E.M.F. round the secondary is proportional to the total change in the number of lines of induction linked with it.

230. Coefficient of Mutual Induction. Let us suppose a unit current is circulating in the primary, then a certain number of lines of induction issue from it and of these some are linked with the secondary. This number will depend on the dimensions and relative position of the two circuits; it is called the **Coefficient of Mutual Induction** of the two circuits. Let us denote it by M; then if we double the current in the primary we double the induction—supposing the field free from iron—everywhere. Thus the number of lines of induction through the secondary is doubled, and if a current i circulates in the primary, the total induction through the secondary is Mi.

In estimating the number of lines of induction linked with the circuit, the number of turns in the circuit must be considered. If M_0 lines pass through each turn and the number of turns is m, then the total number of linkages

---

[1] Glazebrook and Shaw, *Practical Physics.*

is mM_0. If unit current be started in the primary there
will be an E.M.F. M_0 in each of the m turns of the secondary,
thus the total E.M.F. in the secondary is $m \times M_0$.

It can also be shewn that if a unit current circulates in
the secondary, the number of lines of induction which thread
the primary is also M.

Thus we have the following

DEFINITION. *The number of lines of induction due to unit
current in one circuit which thread a second is known as the*
Coefficient of Mutual Induction *between the two.*

It follows from the above that if the coils be moved so that their
coefficient of mutual induction is changed from M to M', and if simultaneously the current in the primary changes from i to i', then the total
electromotive force in the secondary is $Mi - M'i'$ and the total flux of
electricity round it is $(Mi - M'i')/R$, where R is the resistance of the
secondary.

231. Lenz's Law. The law to which reference has
already been made which states that in all cases of electromagnetic induction the induced currents have such a direction
that their reactions tend to stop the motion to which they are
due is known as Lenz's Law.

232. Self-Induction. When a current is started
round a coil of wire the number of lines of induction through
that coil is varied. In consequence an induced electromotive
force tending to oppose the current is produced, the current
rises to its steady value less rapidly than it otherwise would;
some of the energy supplied by the battery goes to establish a
magnetic field in the neighbourhood.

When the current is broken, the number of lines of
induction through the circuit is reduced, an electromotive
force is set up acting in the same direction as the original
electromotive force tending to maintain the original current.

This is called the electromotive force of self-induction.

DEFINITION. *The number of lines of induction due to unit
current in a circuit which are linked with the circuit is called
the* Coefficient of Self-Induction *of the circuit.*

Hence if L be the coefficient of self-induction of a circuit carrying a current i and having a resistance R, then the total E.M.F. of self-induction round the circuit is Li and the total flux of electricity it causes is Li/R.

233. Unit of Inductance. The practical unit in terms of which coefficients of self or mutual induction are measured is called the Henry.

DEFINITION. *The coefficient of mutual induction between two circuits is one* **Henry** *when, if a current of 1 ampere is passed round the primary circuit, the flux of electricity round the secondary of resistance R ohms is that carried by a current of $1/R$ amperes flowing for one second.*

234. Observations on Induction.

EXPERIMENT 56. *To shew the production of the current of self-induction.*

(1)   A battery is connected through a key with a coil of many turns having a large coefficient of self-induction—a coil with an iron core is usually employed.   The ends of the coil, AB, Fig. 236, are also connected through a rough galvanometer.

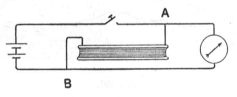

Fig. 236.

When the key is closed a current passes through the coil and the galvanometer traversing both in the direction from A to B suppose; let it deflect the north end of the galvanometer needle to the right and place a stop in contact with the needle so that this deflexion cannot take place.

Break the circuit at the key.   An induced electromotive force acts through the coil from A to B in the same direction

as the original E.M.F.; the induced current now passes through
the galvanometer from B to A and the north end of the needle
is deflected to the left.

(2)   The same effect may be shewn better with the aid of
an incandescent lamp in place of this galvanometer.   The
battery can be adjusted so that the steady current is insufficient
to glow the lamp; on breaking circuit it glows instantaneously,
the much more powerful electromotive force of self-induction
drives sufficient current through it to render it incandescent;
if the current be again made it glows again for an instant;
the E.M.F. of self-induction at make checks the current in the
coil and causes a large flow through the lamp; by arranging
an interrupter in the circuit and working it continuously the
lamp may be kept in a steady state of glow.

EXPERIMENT 57.   *To shew the production of an induced
current due to the motion of a coil in the earth's field.*

Connect a coil resting on the table, Fig. 237, to a somewhat
sensitive galvanometer.   Lift it up quickly and reverse its

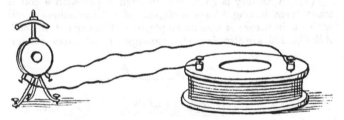

Fig. 237.

position, laying it down with the bottom face uppermost; the
galvanometer shews an instantaneous current.

The lines of magnetic induction due to the earth traverse
the coil in a direction opposite to their previous one; this
change has caused an induced current in the coil.

(3)   Arrange the coil as shewn in Fig. 238, so that it can
rotate about an axis in its own plane.   Turn the coil about
this axis through 180°.   The galvanometer needle is deflected,
but returns to zero again; an induced current has passed;

continue the rotation through a further 180°; the needle is again deflected but in the opposite direction; a second induced current opposite in direction to the first has traversed the circuit; thus by continuing the rotation we get an alternating current in the circuit.

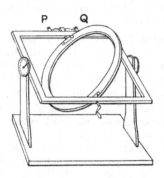

Fig. 238.

If we call X and Y the two faces of the coil and suppose that in the original position a maximum number of lines of magnetic induction due to the earth enter the coil at the face X and leave it at the face Y, as the coil rotates the number of lines entering at X decreases till at last it vanishes. As the rotation continues lines of induction begin to enter at Y increasing up to a maximum when a rotation through 180° has been completed. Throughout this part of the rotation the E.M.F. has the same direction, the lines of force entering X are decreasing, those entering at Y are increasing; and these two effects contribute to the same result; as the rotation continues through a further 180° the reverse is the case, the lines entering Y decrease, those entering X increase; thus the induced current changes its direction, to reverse again when the first position is reached[1].

We can obtain by calculation a relation between the

[1] When the effects of self-induction are considered this statement needs a slight modification which is however not important for our present purpose.

induced current and the strength of the field in which the coil is rotated and hence by measuring the current can calculate the strength of the field. This is made use of in the earth inductor.

By attaching a split ring commutator to the axis of the coil we can divert all the currents into the same direction in the external circuit.

The arrangement has been already described in Section 157 and as adapted to the present purpose is illustrated in Fig. 245 a. The two halves of the split ring are mounted on the axis about which the coil rotates, being insulated from each other, and the ends of the coil are connected to these.

Two springs to which the ends of the external circuit are connected press against the split ring. In one position a current circulating in the coil in a given direction passes round the external circuit from Q to P. As the coil rotates the connexions at the commutator are reversed; the commutator is so fixed to the coil that at the same moment as this occurs the direction of the current is reversed in the coil; hence its direction in the external circuit will still be from Q to P. Thus by the aid of the commutator the currents in the external circuit are all diverted into the same direction.

Such an arrangement of a coil rotating in a magnetic field and thus producing a current constitutes the simplest form of an electromagnetic machine. See Section 240.

235. Arago's Disc. If a conductor such as a copper disc be moved near a magnet, induction currents are set up in the conductor; these since they circulate in closed curves in the substance of the conductor are known as eddy currents. The electromagnetic forces called into play tend to stop the disc; they tend therefore to move the magnet in the same direction as the disc.

This is illustrated in Arago's apparatus, Fig. 239. A copper disc is made to rotate about a vertical axis and a bar magnet is pivoted above the centre of the disc; the magnet is separated from the disc by a sheet of glass which protects it from air currents.

On setting the disc in motion the induced currents produced act on the magnet and drag it round after the disc, its speed gradually increases until it is moving at the same rate as the disc; when this is the case the disc no longer cuts

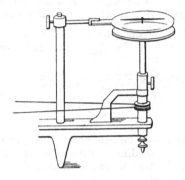

Fig. 239.

lines of induction due to the magnet and the induced currents cease. If the disc be now stopped, with the magnet still moving, induced currents are again started and these act as a drag on the magnet, gradually bringing it to rest.

In a somewhat different form of the experiment the disc is mounted so that it can rotate between the poles of an electromagnet.

Before the electromagnet is excited the disc turns easily; when the magnet is excited induced currents are set up in the disc and the forces called into play tend to stop the motion; much force is needed to keep the disc rotating while the disc becomes strongly heated by the eddy currents which circulate in it.

236. Electromagnetic method of measuring permeability. As we have already seen the total quantity of electricity which circulates round a secondary circuit is equal to the ratio of the change in the number of lines of induction linked with the circuit to the resistance of the circuit. The quantity of electricity can be measured as in Section 174 by means of a ballistic galvanometer included in the secondary circuit; hence by multiplying this by the resistance we can find the total change of induction through the circuit. On this are founded several methods for measuring the permeability of iron or other magnetic materials.

Thus, if the iron take the form of a rod, we may wind a long coil round it as in Fig. 240. This constitutes the

Fig. 240.

primary coil, and the iron can be subjected to a known magnetising force, by sending a current of measured strength through the coil; the current is measured by including an ammeter in the circuit.

A few turns of insulated wire wound round the coil and bar near its centre constitute the secondary coil; these are connected to a suitable ballistic galvanometer; by means of some subsidiary experiments the relation between the deflexion of this galvanometer and the quantity of electricity which circulates in the circuit is determined; thus, when this is done, by observing the throw of the galvanometer needle the total flux of electricity can be found.

If now the primary circuit be closed, the iron is subject to a definite magnetising force, and the galvanometer needle disturbed; the total induction in the iron is determined by calculating from the induction throw the total flux of electricity in the secondary and multiplying this by the resistance. Knowing then the induction and the magnetising force, we can find the permeability. The experiment in this form is open to some objections for poles are formed near the end of the rod and the magnetising force in the rod is modified by their presence. Moreover lines of force pass back through the air round the rod, and since the secondary coil cannot be in contact with the iron some of these lines of force pass through it and modify the result. The total induction linked with the coil is the induction in the iron together with that in the air surrounding the iron.

These difficulties are overcome by forming the iron into a closed ring, and winding the primary coil uniformly round it. In this case the lines of induction lie within the iron while the magnetic force due to a current i is $4n\pi i/l$ where l is the length of the axis of the ring and n the number of turns on it.

The secondary consists as before of a few turns of wire connected to a ballistic galvanometer and the induction is measured by the throw of this instrument.

A rheostat is included in the primary circuit and by means of this the strength of the current and therefore of the magnetising force can be adjusted. The arrangement of

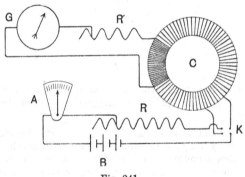

Fig. 241.

apparatus is shewn in Fig. 241 in which C is the ring with the two coils wound on it, B the battery, A an ampere meter for the primary current, R the rheostat and K a reversing key. The induced current is measured by the ballistic galvanometer G.

By means of this arrangement magnetisation and hysteresis curves can be obtained as in Section 216.

The iron is first demagnetised by rapidly reversing and gradually reducing the current. A suitable resistance is taken out of the primary and the current made, the corresponding throw of the galvanometer gives the induction; the iron is again demagnetised, the primary resistance decreased and the current again made, thus a second value of the

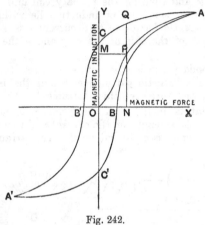

Fig. 242.

induction is found and the curve of induction can be drawn as shewn in Fig. 242 for a magnetising force increasing from zero.

To form the descending part of the curve we start with the iron fully magnetised and suddenly reduce the current by a known amount. The induction throw observed gives us the change in induction, and subtracting this from the induction due to the full current, we can find the point on the descending branch which corresponds to the value to which the current was reduced. The curve thus obtained agrees with that obtained by the magnetometric method.

EXAMPLES ON ELECTROMAGNETISM AND INDUCTION OF CURRENTS.

1. A magnet is suspended at the centre of a coil of wire, the coil being placed in an east and west vertical plane. When there is no current passing, the magnet oscillates 15 times in 10 seconds, and when a current is passing the swings number 25 in 10 seconds. Find the field produced by the current.

$$(H = \cdot 18 \text{ c.g.s. unit.})$$

2. A magnet is placed in a coil with its axis parallel to the axis of the coil. It is then quickly withdrawn. Will the direction of the current in the coil depend upon the direction in which the magnet is drawn out of it?

3. The ends of a circular coil of wire are connected with a galvanometer and the coil can be made to rotate about one of its diameters. On turning the coil half round, the galvanometer needle is momentarily deflected, but on causing it to rotate in one direction continuously and rapidly no effect is produced. Explain these two results.

4. A plane rectangular iron frame is placed vertically so that it faces due magnetic north. It is made to fall northward into a horizontal position. Describe what will be the result.

5. A circular coil of wire spins about a diameter which is placed vertically (1) at the magnetic equator and (2) at the magnetic pole. Explain the nature of the electric currents, if any, which will flow round the coil in each case.

6. Two coils A and B are placed parallel to each other, the former being connected to a battery and key, and the latter to a ballistic galvanometer. Shew that, if the galvanometer has a very high resistance, the throw of the needle on starting a current in A is approximately proportional to the number of turns in B, whilst if the galvanometer resistance is very small the throw is nearly independent of the number of turns.

7. A very long magnet, lying along the axis of a coil of wire, is moved in the direction of its length with a velocity of 10 cm. per second. The pole strength of the magnet is 5 c.g.s. units. Find in volts the greatest value of the E.M.F. induced in the coil, which has a mean radius of 14 cm. and possesses 800 turns.

8. Twenty turns of wire are wound round a square frame, each side of which is 30 cm., and the whole is rotated about a diameter in a magnetic field of 500 c.g.s. units strength. Find the number of revolutions per second in order that the average E.M.F. in the wire may be 1 volt.

CHAPTER XXIV.

APPLICATIONS OF ELECTROMAGNETIC INDUCTION.

237. Principle of Transformers. The phenomena of induction are made use of in induction coils and transformers to enable us to produce from a given current one of a different electromotive force.

Consider a coil of a moderate number of turns of thick insulated wire through which a current can be passed arranged as in Fig. 240 above. This is called the primary coil. Let there be wound over this m times as many coils of thinner wire—the secondary coil. If a current circulate in the primary, a certain number of lines of induction pass through it; these are also linked with the secondary, but since the number of turns on the secondary is m times as great as that on the primary the number of linkages is m times as great; if the primary current be now broken, induced electromotive forces are set up in both coils but the E.M.F. of induction in the secondary will be m times as great as in the primary; thus by breaking the current in the primary we get an induced current of much greater electromotive force in the secondary; if now we make the primary circuit, an induced current of great electromotive force is set up in the opposite direction in the secondary; and by alternately making and breaking the primary alternating induced currents of high electromotive force are produced in the secondary.

The effect is increased by introducing soft iron into the core of the coils for in this way the total induction is greatly increased.

Again the total flux of electricity in the secondary depends on the total change of induction and is therefore the same whether the break in the primary be sudden or prolonged. But the time during which this flux lasts is reduced by making the break a sudden one; thus the strength of the current in the secondary at each instant while it lasts is increased by increasing the suddenness of the break or make.

238. The Induction Coil. The induction coil usually takes the form shewn in Fig. 243. The iron core consists of a bundle of soft iron wires; by separating the iron in this way the eddy currents which would be produced in it, were it in one continuous mass, are greatly reduced.

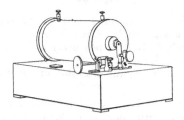

Fig. 243.

Various arrangements are in use to secure the interruption of the primary; the simplest, shewn in Fig. 243 a, consists of a stiff strip of brass or iron which carries a soft-iron hammer at one end, and is placed so that the hammer in the equilibrium position rests near the end of the core. A screw with a platinum point presses against a bit of platinum on the spring. This screw is connected to the battery while the spring itself is in connexion with one end of the coil. The other end of the coil is connected to the second pole of the battery, and in this condition the circuit is complete. The soft iron core is magnetised by the current and attracts the hammer and spring. The contact between the platinum points is thus broken and the current stopped. The core ceases to be magnetised and the spring draws the hammer back again making contact between the platinum points and completes the circuit.

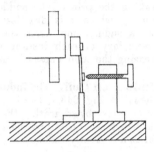

Fig. 243 a.

The condenser is connected between the spring and the screw. When the circuit is broken the energy of the induced current at break, instead of producing a spark between the points and thus drawing out the current, is employed in charging the condenser; the cessation of the current is thus made more rapid.

The secondary coil is wound on over the primary; it consists of a very large number of turns of very fine wire; with a view of preventing too large a potential difference between adjacent turns the wire is wound in sections. By means of such an arrangement the electromotive force of the secondary current is made many times as great as that of the primary and a long spark can thus be obtained from a battery of a few volts.

The quantity of electricity in the secondary current however is not increased in the same ratio, for we have seen that it is equal to Mi/R where R is the resistance of the secondary. Now R is strictly and M approximately proportional to the number of turns of the secondary so that the ratio M/R is not increased by putting more turns on to the secondary.

An induction coil is usually fitted with a commutator such as has been described in Section 157, Fig. 151, in order to reverse the direction of the primary current.

In the induction coil a sudden break is secured by placing a condenser in parallel with the coil in such a way that when the circuit is broken the induced primary current of self-induction which without the condenser would form a small arc across the points at which the break occurs is used in charging the condenser ; thus the cessation of the current is made more rapid than without the condenser, and the whole of the secondary flux is concentrated into a shorter time. The secondary current is increased. Some recent experiments of Lord Rayleigh have shewn that the same effect can be produced without the use of a condenser by making the break sufficiently sudden while if the condenser be too large its action may reduce the secondary current.

239. Transformers. A machine employed for the purpose of changing the electromotive force of an alternating current is called a transformer.

The induction coil and the arrangement described in Section 234 constitute transformers; if the number of turns in the secondary be m times as great as in the primary then the secondary E.M.F. is approximately m times as great as that of the primary. In a transformer m may be greater or less than unity; in the first case as in the induction coil the E.M.F. is transformed upwards by the machine; in the latter it is transformed downwards ; the E.M.F. in the secondary is less than that in the primary.

In order to cause as many as possible of the lines of induction due to the primary circuit to pass through the secondary, the iron circuit of a transformer is usually completely closed, forming a ring round which the two wire circuits are wound, as shewn in Fig. 244.

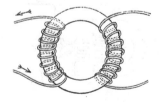

Fig. 244.

In electric light and power installations transformers are usually employed to reduce the volts of a supply system from the value which has been found suitable for transmission from the central station to that suited to the consumers' lamps or motors. The consumer needs a certain supply of energy. The supply of this per unit of time is measured by the product EC of the volts and amperes. To supply a large current however requires a large amount of copper in the leads otherwise the loss due to heating becomes too great; it is therefore in many cases economical to supply a small current at a high voltage. The consumer may require for his work a larger current but at a less voltage; this can be given him by means of a step-down transformer.

In some cases two transformers· may be employed, *e.g.* in working a large tramcar system or on a railway. The machine produces the electricity at a moderate voltage. It is transformed up in order that copper may be economised in the leads which carry it to the more distant parts of the system. At these distant points it is transformed down so as to be conveniently supplied to the motors of the cars.

240. Electromagnetic Machines. We have seen in Section 234 how a supply of electricity can be obtained by suitably rotating a circuit in a magnetic field. The experiment there described contains the germ of the many dynamos and electromagnetic generators in existence. In its simplest form such a machine would consist of a single elongated coil of a few turns mounted as shewn in Fig. 245 *a* between the poles of a magnet. This coil can be made to rotate about a horizontal axis lying in the plane of the paper; its ends are connected to a split ring commutator giving direct current in

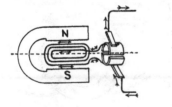

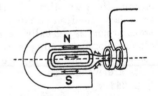

Fig. 245 *a*.   Fig. 245 *b*.

the external circuit or, for an alternate current machine, to two slip rings, 245 *b*. The rotating coil is known as the armature of the machine. As the coil rotates the E.M.F. changes from zero when the plane of the coil is perpendicular to the lines of force to a maximum when it is parallel to them.

241. Magneto-electric Machines. The small magneto-electric machines, Fig. 246, used sometimes in

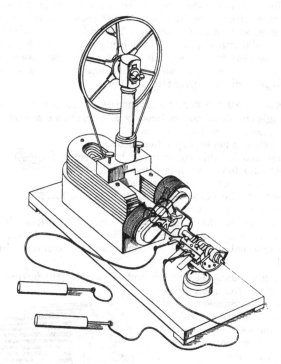

Fig. 246.

medical practice act on the same principle. Two coils of wire with parallel iron cores are connected in series and arranged so that they can rotate about a common axis parallel to the cores. The rotation takes place in front of a horseshoe magnet so that the ends of the cores come alternately opposite to the poles of this magnet. The other extremities of the cores are connected by a bar of soft iron. Thus in one position of the coils lines of induction run from the north pole of the magnet through the axis of one coil and return to the south pole of the magnet through the axis of the other; on turning the coils through 180° the direction in which the lines of induction are linked with them is reversed. Thus an electromotive force is produced round the coils which is reversed at each half rotation. The ends of the coils come down to slip rings against which brushes rub, and an alternating current is produced in an external circuit connected with the brushes. By using a commutator instead of two slip rings a direct current is obtained.

242. Shuttle wound Armature. Returning now to the single coil described in Section 240, imagine a second coil to be fixed, as in Fig. 247, on the same shaft in a plane perpendicular to the first and let this commutator be divided into four segments—two for each coil — instead of into two. Further let the coils be connected to the segments so that each in turn

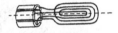

Fig. 247.

is connected to the brushes when near the position for a maximum electromotive force. The fluctuations of the voltage in the external circuit will be reduced and the average voltage raised.

By continuing the process and increasing still further the number of the sections we arrive at the Siemens shuttle wound armature in which a large number of coils are wound on a central boss of laminated[1] soft iron, each coil being connected to two sections of the commutator, the brushes of which are arranged so as to tap off the current from each section in turn as the E.M.F. in that

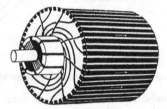

Fig. 248.

---

[1] The iron is made of a number of thin laminæ to reduce eddy currents.

section reaches its maximum value. In each case the end of one coil and the beginning of the next are connected to the same section of the commutator so that the various coils on the armature form a continuous circuit. Fig. 248 shews such an armature.

This armature rotates between the poles of a magnet shaped so as to come as close as possible to the wire of the coils and thus increase the induction through them.

243. Dynamo Machines. To induce the current a permanent magnet may be used, but in modern machines an electromagnet is ordinarily employed. The coils of this magnet are spoken of as the field coils of the machine. The field coils may be supplied with current from some external source; the machine is then said to be separately excited.

Separate excitation however is not always required, for suppose the field coils are connected up in series with the armature through the brushes and the machine started. Sufficient traces of magnetisation are usually left in the iron of the machine to start a small current; this passing round the field coils magnetises them more strongly, thus increasing the current which again reacts on the field, the machine is then self-excited. Such a machine is spoken of as a Dynamo Machine.

The volts produced by a dynamo depend on its construction and on the nature of the iron used in it, they also depend on the speed being, when the magnetisation of the iron approaches saturation, approximately proportional to the speed. In this case the strength of the field is nearly constant, and the number of lines of induction cut in a given time by the armature will be proportional to the speed.

For some purposes the coils of a dynamo are connected so that the whole current through the armature passes through the external circuit, and the field coils in series. The machine is said to be series wound, Fig. 249. Since the whole current has to traverse the field coils, these consist of a few turns of thick wire.

In this case if the resistance in the external circuit decreases the current in the armature and field coils increases.

The increase of the current in the field coils produces for constant speed a rise in the E.M.F. of the machine which tends to raise the external current still further, thus the machine is not suitable for steady working in a circuit of variable resistance, *e.g.* in an incandescent lamp circuit where the resistance depends on the number of lamps in use.

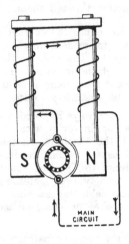

Fig. 249.   Fig. 250.

In a shunt wound machine, Fig. 250, the field coils consist of a large number of turns of thin wire, the ends of which are connected to the brushes. Thus the field coils and the external circuit are in parallel.

If for such a machine the resistance in the external circuit falls, thus causing the machine for a given voltage to give an increased current, a smaller fraction of that greater current circulates round the field coils; thus the strength of the field does not necessarily increase as the current in the armature increases and the voltage of the machine remains more nearly constant for varying currents than is the case in a series machine.

The rise or fall in voltage depends on the relation between the various resistances which will determine whether the rise produced in the current in the field coils is greater or less than the fall due to the fact that a smaller fraction of that current circulates through them.

In a compound machine, Fig. 251, some of the field coils are in series with the external circuit as in a series machine, and in addition to this a number of turns of thin wire are wound on in parallel with the external circuit.

When the external resistance is reduced the current in the armature rises; so far as the series coils are concerned the field increases; a less fraction, however, of the whole current circulates through the shunt coils and the resistances are arranged (1) so that

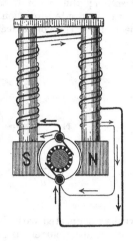

Fig. 251.

the field in consequence is reduced, and (2) so that the reduction just counterbalances the rise due to the action of the series coils: thus the voltage remains constant.

244. The Gramme ring. Another form of armature is the Gramme ring. Let us suppose we have a field of magnetic force in which the lines are parallel to the paper, as shewn in Fig. 252. Consider a coil whose plane is perpendicular to the paper, and let it move about an axis parallel to its own plane and perpendicular to the paper. When the coil is in the position A, a maximum number of lines of force pass through it. As it rotates into position B, the number decreases until none pass through; from B to C the decrease continues, the lines of force again become linked with the coil, but in the negative direction opposite to that in which they previously passed through, and at C the number of negative lines is a maximum. Thus during the motion from A to C an E.M.F. acts on the coil and a current circulates in it. After passing C the negative lines decrease in number to vanish at D when lines begin to traverse the coil in the positive direction; thus from C to A an E.M.F. acts round the coil in the opposite direction to that previously existing. If the ends of the coil be carried to two rings on the axis and

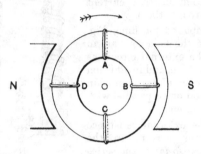

Fig. 252.

brushes, connected with the external circuit, rub against these an alternating current is produced in this circuit; by using a commutator instead of the rings, the current may be made to traverse the external circuit always in the same direction. The E.M.F. in the circuit varies as the coil rotates, and is greatest when the coil is in position B or D.

If now, instead of a single coil we have a series of such coils arranged so that their centres lie on a circle with its centre in the axis of rotation, and corresponding to each have two segments of a commutator set so as to pick off the current from each coil when the E.M.F. in that coil is a maximum, we increase the current greatly.

If further the coils be all wound about a core of iron wires shaped like an anchor ring the magnetic induction and the corresponding E.M.F. will be greatly increased.

Moreover the coils may be all continuous as in the diagrammatic illustration of Fig. 253.

A continuous length of wire is wound round the ring-shaped core and from a number of equidistant points round the ring wires are led to the respective sections of the commutator.

Brushes make contact at E and E' respectively.

Owing to electromagnetic induction all the turns of the armature in the half ABC tend to produce a current traversing the armature in the direction ABC, those in the half CDA tend to produce a current in direction ADC. These two currents, each of strength $I/2$ let us suppose, combine to produce in the external circuit a current of strength I.

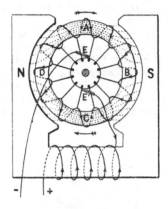

Fig. 253.

We notice that since the voltage of the machine between E and E' is that due to all the turns on the half ABC it may be considerable; the resistance, however, of the armature, being that due to all the turns, is also considerable; hence the machine is not suited to give a very large current.

As in the shuttle wound form of the dynamo the field magnets are usually electromagnets, the current being started by the residual magnetism of the iron, the machine being either series, shunt or compound wound. In the figure a series wound machine is shewn.

245. Armature Reactions. It should be noted that the current in the armature produces its own field of magnetic force, hence the lines of induction through the armature are not solely those due to the field magnets, and the theory is more complex than that given above; besides, since the current round the armature varies continually, effects of self-induction have to be considered; in consequence, in an actual machine when working, the brushes have not the position shewn but are displaced in the direction of motion.

246. Alternate Current Machines. Modern alternate current machines are more complex than that figured in Fig. 245 b—a simple coil connected to two slip rings. Thus the field magnets of a Siemens machine consist of two sets of electromagnets. The ends of the axes of each set are arranged in a circle, the axes themselves being perpendicular to the plane of the circle and the poles being alternately positive and negative. The two series of poles are placed with their planes parallel and at a short distance apart in such a position that the north poles of one set are opposite to the south poles of the other and *vice versâ*, as in Fig. 254. Thus between the

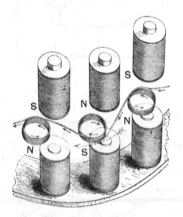

Fig. 254.

two there is a field of force; and the lines of force will run alternately upwards and downwards.

The armature revolves in the gap between the two series of

magnets. It consists of a series of coils, the same in number as the magnets, placed round the circumference of a disc, with their axes parallel to those of the magnets, in such a way that their centres are equally distributed round a circle of the same diameter as that on which the poles of the magnets lie.

The windings of the alternate coils are opposite in direction, the wire is continuous round all the coils and its two ends are connected to slip rings on the axle. From these the current is conveyed by means of brushes to the external circuit. Now start from a position in which the centres of the armature coils lie on the axes of the magnets. In one-half the coils, as the machine rotates, the number of lines of force which pass through from below upwards is increased, in the other half it is decreased. Thus the electromotive force round the alternate coils of the armature is opposite; but the direction of the windings round alternate coils is also opposite, hence the electromotive force retains everywhere the same direction through the wire of the armature. When the armature is moved the E.M.F. starts from zero, gradually rises and falls again to zero as the armature coils again come opposite to the magnet coils. As the motion continues the same process is repeated, but the E.M.F., as the armature coils move one section forward, is opposite in direction to that in the previous section; an alternating current is produced, it passes through the value zero each time the moving poles of the armature come opposite the fixed poles of the magnets and has opposite signs between each consecutive pair of such positions.

There are various other forms of alternate current machines, for which reference should be had to special books on the subject.

See for example S. P. Thompson's *Dynamo Electric Machinery*.

In some of these the armature is fixed while the field magnets rotate, the advantage of this is that a commutator is not then required in a moving part of the circuit in which the alternate current is circulating.

It should be noted, moreover, that the current does not alternate in the field magnets; in some cases permanent magnets are used, in others a small direct current machine is employed to excite them.

247. Electric Motors.

In a dynamo machine mechanical energy is transformed into electrical. The armature is made to rotate in a magnetic field and a current is formed. This action is in all cases, with direct current machines, reversible; if we supply a current the armature will rotate, the dynamo becomes a motor; electrical energy can by its aid be transformed into mechanical. There will be as many forms of motors as there are of dynamos.

When a given E.M.F. is first applied to the terminals of a motor at rest the resistance is small, a large current passes and the machine begins to rotate. This rotation sets up an E.M.F. in the machine opposite to that driving it, and the current is reduced; if the driven machine is doing no work and if there be no friction the speed will go on increasing until this opposing E.M.F. is equal to that applied to the machine; the current then vanishes.

248. Transformations of Energy in a Motor.
Now let us call E the impressed E.M.F. and e the back E.M.F. of the motor. If the machine is doing work the back E.M.F. of the motor is less than the impressed E.M.F. and the current is $(E-e)/R$, where R is the resistance of the circuit. If we call this current I, the energy supplied per unit time to the machine from outside is EI, the energy used by the motor in doing external work is eI, and the energy lost as heat in the armature is $(E-e)I$, or writing in the value of I we have

Energy supplied to system $E(E-e)/R$,

Energy transformed by motor $e(E-e)/R$,

Energy lost as heat $(E-e)^2/R$.

We can represent the first two of these quantities in a diagram thus, Fig. 255.

Let $ABCD$ be a square of which the side AB represents E.

Take AK and AM each equal to e, and draw MON and LOK parallel to AB and BC.

Then KB and MD are each equal to $E-e$, and the parallelogram LB is equal to $E(E-e)$ and therefore represents the energy supplied, while the parallelogram NK is $e(E-e)$ and represents the work done by the motor.

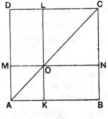

Fig. 255.

As e increases, that is as K moves along towards B, the area of this parallelogram increases at first and then decreases, reaching its maximum value when K is midway between A and B so that e is equal to $E/2$. In this case the external work done is a maximum.

Thus a motor does the greatest amount of work when its speed is such that the back E.M.F. is half the impressed. The current then is $\frac{1}{2}E/R$, it has half the value it would have if the motor were at rest.

By the efficiency of a motor is meant the ratio of the work it does to the energy supplied. Thus the efficiency is measured by $e(E-e)/E(E-e)$ or e/E; it is greatest then if $e = E$. In this case the motor takes no energy from the source and does no work. A distinction then has to be drawn between the condition for maximum efficiency and that for maximum work done; in the latter case, since $e = \frac{1}{2}E$ the efficiency is only one-half.

Since the energy lost as heat is $(E-e)^2/R$ we notice that when the external work is a maximum and e is $\frac{1}{2}E$ its value is $\frac{1}{4}E^2/R$. Under these conditions the external work $e(E-e)/R$ is also $\frac{1}{4}E^2/R$ while the total energy supplied is $\frac{1}{2}E^2/R$. Thus when the external work is a maximum it is equal to half the energy supplied.

249. Starting a Motor. The back electromotive force in a motor does not reach its full value until the motor has acquired its final speed; accordingly the current taken by a motor at starting is greater than that which it ultimately requires to drive it; indeed, if the full voltage were applied direct to the armature the current would be too great, and the armature would be destroyed. For this reason a starting resistance is usually employed in series with the armature of the motor; the volts are applied at first through this resistance which is gradually cut out of the circuit as the speed rises, until finally the full pressure of the supply is on the terminals of the machine.

EXAMPLES ON ELECTROMAGNETISM AND ELECTROMAGNETIC INDUCTION.

1. A single turn of wire coiled into the form of a circle of 12 cm. radius is placed in the magnetic meridian, and two circular turns of wire of radius 24 cm. are placed so that their common plane and centre coincide with that of the first circle. A small magnet is suspended horizontally at the centre. What effect, if any, will be produced on the magnet if the same current be sent in one direction through the single wire, and in the other direction through the double wire?

2. The single wire mentioned above (see Question 1) is turned about a vertical axis until in a plane perpendicular to its former position, and the new position of the needle is observed. The direction of the current in the single wire is then reversed; calculate the deflexion of the needle produced by the reversal from the following data:

Current flowing through the coils $= 10$ amperes,

$H = \cdot 18$ c.g.s. unit.

3. A current of 10 amperes flows along an indefinitely long straight wire; find the force which it will exert upon a magnetic pole of strength 20 units placed at a distance of 6 cm. from the wire.

4. A wire is bent into the form of a circle of 10 cm. radius, and a current of 2 amperes is passed through it. Find the force exerted by the current at a point on the axis and at a distance of 20 cm. from the circumference.

5. Two infinitely long straight wires are placed parallel to one another, and currents of 5 and 8 amperes respectively are passed through them in the same direction. Find (1) the magnitude, (2) the direction of the force acting upon unit length of either wire.

6. A platinum wire is hung from a loop in a copper wire so that its lower end just dips into a vessel containing mercury, and the copper wire and the mercury are connected to the opposite poles of a battery. Describe and explain the movements that will take place.

7. A large circular coil is in series with a smaller one and a ballistic galvanometer. When the larger coil is quickly turned through 180° about a vertical axis in the Earth's field, a deflexion of 30 divisions results. The smaller coil when similarly turned between the poles of a horse-shoe magnet produces a deflexion of 70 divisions. Calculate the value of the field in which the smaller coil is turned. The following are the necessary data:

Number of turns on large coil $= 80$.

„   „   „   small „ $= 10$.

Radius of large coil $= 50$ cms.

„   „ small „ $= 4$ „

$H = \cdot 18$ c.g.s. units.

8. Two horizontal brass rods are placed parallel and at a distance of 1 metre apart. A third rod slides over them parallel to itself with a uniform velocity of 10 metres per second. Find in volts the E.M.F. between the ends of the fixed rods—assuming the earth's vertical magnetic force to be ·47 C.G.S. unit.

9. A hoop of copper is set rotating about a diameter as an axis. It is placed in a magnetic field with its axis of rotation (1) parallel, (2) perpendicular to lines of magnetic force. If the mechanical friction and the initial velocity of rotation are in both cases the same, explain why it comes to rest in a shorter time in one position than when in the other.

10. A bar of soft iron is thrust into the interior of a coil of wire, whose terminals are connected to a galvanometer. Could the coil and bar be placed in such a position that no induced current might pass through the galvanometer?

11. The poles of a battery are connected in turn by

   (1) a long straight insulated wire,

   (2) the same wire coiled into a close spiral,

   (3) ,,   ,,   ,,   ,,   round a soft iron core.

Describe and discuss what happens in each case on breaking the circuit.

CHAPTER XXV.

TELEGRAPHY AND TELEPHONY.

250. The Electric Telegraph. In the electric telegraph the current is used to transmit a signal to a distance. In its simplest form a wire uniting the two places and carried on insulating supports is connected at the transmitting end, *A*, Fig. 256, through a key with one pole of a

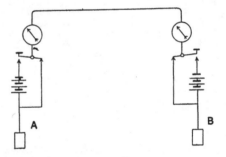

Fig. 256.

battery; the other pole of the battery is put to earth by being connected to a large metal plate sunk in the ground. At the distant station, *B*, the wire is joined to one terminal of a galvanometer of which the second terminal is put to earth in the same manner. On depressing the key a current passes through the wire deflecting the galvanometer needle, and returns by the earth to the sending station, and thus signals can be transmitted from A to B. The needle of the galvanometer usually swings in a vertical plane, its motions are indicated by a pointer outside the case of the instrument. By means of a double key at A either end of the battery can be connected at will to the line; thus the direction of the currents and therefore the direction of the deflexion in the receiving instrument can be reversed. Each letter of the alphabet is represented by a suitable combination of deflexions to right and left.

By adding a second receiver at A and a second battery and key at B, connected as shewn in Fig. 256, the above line can be made to work either way. With the keys in their normal positions both ends of the lines are earthed, and as shewn in the figure the negative poles of both batteries are to earth, the positives being insulated. When the key A is depressed the current passes and is registered on the receiver at B.

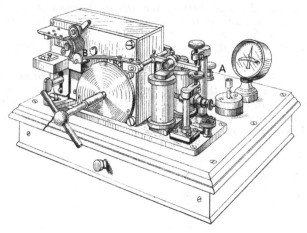

Fig. 257.

251. The Morse Instrument. In the Morse receiver, Fig. 257, a narrow continuous strip of paper is drawn by clockwork from a drum between rollers.

A lever, AB, with a pen or style at the end B carries

a piece of soft iron at A; the current from the distant station can pass round an electromagnet below A; when this is the case the iron is attracted and the pen drawn up against the moving paper; when this circuit is broken a spring draws the pen away again; thus a dot or a dash is marked on the paper depending upon the length of time during which the contact is maintained.

The letters are represented by a combination of dots or short strokes and dashes or long strokes as shewn in the next section.

In practice the current from the sending station would probably not be strong enough to move the lever of the Morse instrument, it is therefore used to work a relay instead.

This instrument consists of a very light lever actuated by an electromagnet through which the distant current passes. When the lever is attracted by the electromagnet it closes the circuit of a local battery through the Morse instrument. Since this current has only to traverse the coils of the instrument and the wires connecting them, instead of the many miles of the line, it can easily be made of sufficient strength to work the instrument.

The Morse key which is used as a sending instrument is shewn in Fig. 170, the line is connected to the fulcrum of the lever, and in the normal position of the key is in circuit with the relay; on depressing the fulcrum this circuit is broken and connexion is made between the line and the battery.

252.　The Morse Alphabet.

A	· —	K	— · —	U	· · —
B	— · · ·	L	· — · ·	V	· · · —
C	— · — ·	M	— —	W	· — —
D	— · ·	N	— ·	X	— · · —
E	·	O	— — —	Y	— · — —
F	· · — ·	P	· — — ·	Z	— — · ·
G	— — ·	Q	— — · —		
H	· · · ·	R	· — ·		
I	· ·	S	· · ·		
J	· — — —	T	—		

253.　The Telephone.　In a telephone use is made of the currents produced when a piece of magnetised soft iron

in the form of a thin diaphragm is made to vibrate in front of a coil of wire.

Bell's telephone used originally as a receiver and a transmitter took the form shewn in Fig. 258. A coil of wire

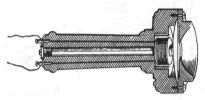

Fig. 258.

is wound on a flat bobbin which surrounds one pole of a permanent magnet, the ends of the coil come to binding screws connected to the external circuit. A disc of thin sheet-iron is supported at its edges very close to the pole of the magnet which carries the coil. This constitutes the transmitter. At the distant station a similar apparatus is connected to the circuit. On speaking into the transmitting instrument the magnetised disc is set into vibration, thus producing induction currents in the coil; these traverse the coil of the distant instrument or receiver, thus setting it into a similar state of vibration and causing it to emit the same sounds as those to which the motion of the transmitter was due.

254. The Microphone. The currents obtained from Bell's instrument used as a transmitter are very weak, and hence for this purpose it has been displaced by instruments working on the principle of the microphone.

If in a circuit containing a battery and a telephone there be a loose contact, as in Fig. 259, in which the vertical rod rests loosely in its supports, and this contact be set into vibration by any means, the variations of the resistance at the contact are so great that corresponding vibrations are produced in the telephone and a sound is emitted; the vibrations

thus produced may exceed in amplitude those of the original
sound and hence it may be intensified.

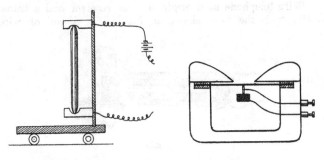

Fig. 259.   Fig. 260.

Various forms of transmitter based on this principle are
in use; in the Blake instrument, Fig. 260, a platinum point,
which is mounted on a disc of elastic material, presses lightly
on a piece of carbon; the current from a battery, usually a
few Leclanché cells, passes through this contact and round
a Bell telephone at the receiving station. On speaking into
the transmitter the vibrations of the disc cause variations in
the resistance at the point and hence in the current.   These
set the receiving disc into vibration and the words are
reproduced.

When the receiver is on its hook the current through the
telephone is broken, the line wire is connected to an electric
bell at the receiving station; this bell can be rung by
pressing a button at the transmitter.   On taking the telephone
off its hook a spring causes the hook to rise, thus breaking the
bell circuit and closing that of the telephone.

CHAPTER XXVI.

ELECTRIC WAVES.

255. Electric Inertia. In dealing with the motion of matter[1] we considered at some length the phenomena of inertia, that property of matter which makes it persevere in the state of rest or motion in which it finds itself, so long as it is not acted on from without.

The electric current also apparently possesses this property of inertia. We have seen that when electromotive force is applied to a circuit the current does not at once reach its final value, there is at the start an induced opposing current of self-induction, while when the electromotive force is removed the current of self-induction at break continues to flow for a brief period. We may compare this to the motion of a body which is resisted by some kind of frictional force depending on the velocity.

When force is applied to such a body it begins to move, at first slowly in consequence of its inertia, then more rapidly until the frictional force is sufficient to balance the impressed force, when a state of uniform motion is reached which continues as long as the force acts; if the force ceases to act the body continues to move for a time in consequence of its inertia, but is shortly brought to rest by the friction.

We might look upon the body as moving all the time with constant velocity so long as the external force acts, balancing the friction, but as having superimposed on this, at the start, a velocity in the opposite direction, and at the finish

---

[1] *Dynamics*, Section 77.

a velocity in the same direction as that of motion; both these die away, both arise from the inertia of the body. At the start the force is storing up kinetic energy in the body, at the finish this kinetic energy carries the body on, and enables it to do work against friction.

In the same way we may look upon the effects of self-induction in an electric circuit as inertia effects. When the electromotive force is applied to the circuit it takes time to overcome the electric inertia of the system, and during this period the system is gaining kinetic energy; when the electro-motive force ceases this kinetic energy is employed in carrying the current on against the resistance of the circuit; thus we may look upon electricity in motion as having kinetic energy.

Again, a charged body possesses potential energy measured as we have seen by half the product of its charge and its potential.

Thus any charged electrical system resembles a dynamical system in being capable of having both potential and kinetic energy. If for example the system be a charged condenser, and the plates be connected by a wire, the potential energy of the charge becomes kinetic energy in the current of discharge, and is transformed into heat by the resistance of the wire. We wish to study these changes a little more fully.

Suppose we have a smooth ball, Fig. 261, resting in a groove on a horizontal table, and attached at opposite ends of a diameter to two springs or pieces of elastic, which we will suppose are originally slightly stretched, and have their ends fixed to the opposite ends of the groove.

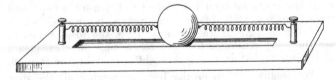

Fig. 261.

Displace the ball in the groove so as to stretch one of the springs further, relaxing at the same time the other. On

releasing the ball it will, if the groove be smooth, move back
to and past its equilibrium position, stretching the spring
which was slack and relaxing the other; it will then come
to rest for a moment, and then reversing its motion move
back through the equilibrium position towards the point from
which it started. A series of oscillations is thus set up which
would go on continuously if the system were quite free from
frictional resistance, but which in reality will die down at a
rate depending on the friction. The ball before being released
has potential energy; this is transformed into kinetic, and
again back into potential.

Consider a ball hanging from the end of a fine vertical
thread, Fig. 262. Raise the ball slightly, still keeping the
thread stretched; in its raised posi-
tion it has potential energy. Then
release the ball; it swings down,
gaining kinetic energy but losing
potential, through its position of
equilibrium, where for a moment
it has its maximum kinetic energy,
rising on the other side, gaining
potential and losing kinetic energy.
This goes on for some time in the
open air, and would continue for
ever if we could entirely remove
the frictional resistance caused by
the air. If now we try to repeat
the same experiment in water, the number of oscillations of the
ball before it comes to rest is greatly reduced, while probably
in oil or some other viscous medium it will not oscillate
at all but sink slowly to its equilibrium position.

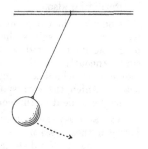

Fig. 262.

A charged condenser corresponds to the raised ball in
having potential energy. If we connect the condenser plates
by a wire we allow a current to flow, the energy takes the
kinetic form, and just as with the ball, if the frictional
resistance be small, we obtain oscillations, so by sufficiently
reducing the resistance of the electrical system we obtain
oscillations of electricity. The positive charge of the condenser passes through the wire from one plate to the other,
the plate which was positive becoming negative, and *vice versâ*,

and this may continue for some time; the current in the wire starts from zero, rises to a maximum, and then dies down to zero, after which it starts again in the opposite direction; and these alternations may go on for some time.

If however the resistance of the wire be very large the number of oscillations will be very small, in fact there may be none, the condenser is discharged at once without any alternations of sign in its plates.

256. Transmission of Magnetic Force. Now in the neighbourhood of any conductor carrying a current there is a magnetic field which is proportional to the current; and when the current is an alternating one the field at each point is alternating also.

Consider then a point at a distance from the conductor; we may ask ourselves the question how are changes in the magnetic field related to those of the current; do they occur simultaneously, or is there a definite interval between the moment at which a change takes place in the current and that at which the corresponding change occurs in the electric or magnetic field at the point P?

It has been shewn that there is a definite interval of time between these two moments, and further that this interval is proportional to the distance between the current and P. The effect moves with a uniform speed.

The changes in the electric and magnetic force are propagated outward from the alternating current and travel through space with a definite velocity.

257. Electric Waves. Moreover experiment has shewn that this velocity of the electric waves is the same as that of light, about 300,000 kilometres or 186,000 miles per second.

The period of the waves, the rapidity with which they succeed each other, will be the same as the period of the alternations of the current, and will depend on the form and dimensions of the electric circuit. Lord Kelvin and von Helmholtz both shewed that if L measures the coefficient of self-induction of the circuit, and C the capacity of the condenser, then the period of oscillation, provided the resistance is very small, is equal to the value $2\pi\sqrt{LC}$.

The theory that changes in electric and magnetic force are propagated through space with a velocity equal to that of light is due to Clerk Maxwell, while Hertz was the first to verify the theory by direct experiment.

To do this Hertz had first to produce oscillations, and then to observe the electric or magnetic changes which take place at a distance. To excite the electric waves Hertz used two square plates, Fig. 263, each 40 cm. in edge, connected

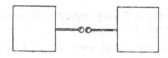

Fig. 263.

by wires about 30 cm. long to two small spherical knobs. These were gilt and placed some 2 or 3 cm. apart. The plates constitute the condenser, the wire the circuit in which the current flows. The condenser is charged by being connected to an induction coil or an electrical machine. When its potential has risen sufficiently the insulation between the knobs breaks down, and an alternating discharge takes place; in Hertz's arrangements the period of the alternation was about $1 \cdot 851 \times 10^{-8}$ seconds, that is to say, if the alternations continued regularly there would be about 50 millions of them in a second; in reality only a few of the alternations pass, they are rapidly damped out, the spark across the knobs lasts for a very short time, and then the plates are again charged up by the machine.

To detect the changes in the electric force at a distance Hertz made use of the principle of resonance.

If we consider any mechanical system which can vibrate in a definite period, then if a series of small impulses having that same period be applied to it, the effect is much greater than that due to a series of much larger impulses which have some different period.

Hertz's receiver, Fig. 264, took the form of a wire circle 35 centimetres in radius, having on its ends two small knobs brought very close together; the period of electrical oscillations in such a wire can be shewn to be the same as that in the smaller circuit. Very small electrical impulses of this period applied to the wire set up oscillations, and sparks which can be observed pass across the gap.

Fig. 264.

Thus Hertz was able to set up electrical waves which travelled out from his emitter with the velocity of light, and to detect them at a distance with his receiver.

258. Wireless Telegraphy. These experiments form the basis of wireless telegraphy.

Instead of Hertz's ring, some form of coherer is usually employed as a receiver.

It was shewn first by Branly that the electric resistance of a glass tube filled with iron filings is very much reduced if electric oscillations fall on it. If the tube be tapped or slightly shaken it recovers its greater resistance.

When used as a receiver by Lodge it was connected up in series with a cell and a galvanometer, Fig. 265; in its normal

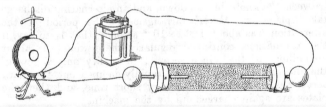

Fig. 265.

condition the coherer resistance is so large that the current passing is very small. When electric waves fall on it the resistance falls, and the galvanometer needle is deflected; the sensitive condition is restored by tapping the board on which the coherer rests.

Wireless telegraphy as now practised is a consequence of the results of these and similar experiments. An oscillator usually of the form devised by Righi is employed. In Fig. 266 A and B are two small spheres or cylinders placed close together on an insulating support. D and E are two other spheres placed near them, and connected to the secondary terminals of an induction coil or electrical machine C. On working the machine a spark passes from D to E through the spheres A and B and oscillations are set up which depend on these spheres.

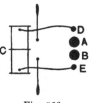

Fig. 266.

In order to concentrate the waves at the receiving station on to the coherer or other receiver a tall pole carrying a wire is employed. The current through the coherer works a relay which drives a Morse instrument and thus signals can be received and read; an electromagnetic arrangement is employed to tap the coherer at frequent intervals and thus keep it in the sensitive state.

In his most recent work M. Marconi has employed a different form of receiver.

The energy needed to transmit electric waves to great distances is very large, hence both the transmitting and receiving apparatus are of necessity on a very great scale.

CHAPTER XXVII.

TRANSFERENCE OF ELECTRICITY THROUGH GASES; CORPUSCLES AND ELECTRONS.

259. Electric Discharge through Gases. The passage of electricity through highly rarefied gases has afforded a field for a very large amount of investigation. The gas is usually contained in a glass tube, Figure 267, into the two ends

Fig. 267.

of which platinum wires are sealed. These can be connected to the secondary terminals of an induction coil, or to a battery of a large number of cells, and the consequences of varying the nature or the pressure of the gas in the tube examined.

It is convenient for some purposes to connect the terminals also to two insulated metal balls outside the tube; by varying the distance between these balls the spark may be made to pass at will, either through the tube, or through the air between the balls, and the distance between the balls at which it just ceases to pass through the air and begins to pass through the tube affords a measure of the resistance offered by the gas in the tube to the passage of the spark.

For many purposes it is best to employ as the negative electrode or kathode a small flat plate of aluminium. The tube is connected to a mercury pump in such a way that the gas it contains can be reduced to a high degree of rarefaction.

At first when the pressure is high the current does not pass through the tube, as the pressure is reduced it begins to pass, and a narrow luminous line is seen down the centre of the tube; the particles of gas along the axis are intensely heated by the passage of the current and become luminous. On reducing the pressure further the glow appears to fill the whole tube. When the pressure is still further reduced to that due to from $\frac{1}{3}$ to $\frac{1}{2}$ of a millimetre of mercury further changes take place. The luminous column is broken up near the positive electrode into a series of portions separated by dark striæ; immediately round the negative electrode is a soft luminous glow, and between it and the striated column a comparatively dark space known as Faraday's dark space; careful observation shews that this negative glow is separated from the kathode by a second dark space at present very narrow.

Carry the exhaustion further, the column of striæ contracts and is followed by Faraday's dark space, the negative glow extends out into the tube, and the dark line between it and the kathode widens, becoming what is known as Crookes' dark space. On carrying the exhaustion still further Crookes' dark space fills the tube, the negative glow is confined to a small space near the anode; the centre of the tube is dark but the surface of the tube glows with a phosphorescent light, the colour of which depends on the nature of the glass, with lead glass it is blue, with soda glass it is green.

260. Crookes' Tubes. The phenomena which occur in the tube in this condition were first investigated by Crookes.

He was led to the belief that in the highly rarefied condition within the tube particles of matter charged with negative electricity were shot off from the neighbourhood of the kathode at a great velocity and travelled in straight lines through the tube.

Thus if an object such as a cross cut out of any material be placed near one end of a tube, as in Fig. 268, usually pear-shaped as in the figure, with the kathode at the narrow end

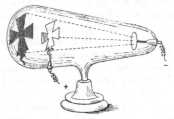

Fig. 268.

a shadow of the cross is projected on to the thick end of the tube. Again Crookes mounted in the tube a small wheel with vanes, arranging it so that the stream from the kathode—the kathode rays—might strike the vanes on one side of the axis only when the wheel was set in rapid rotation.

If the kathode is made concave so as to cause the rays which leave it at each point in the direction of the normal at that point to converge, then a small body placed at the focus or point of convergence is greatly heated; the temperature can by this means be raised sufficiently high to melt platinum.

Various substances, such as some of the rare earths, placed within the tube glow with most brilliant phosphorescence where the rays strike them.

261. Kathode Rays. The kathode rays constitute a current of negative electricity streaming from the negative electrode, and the above phenomena shew that there is a stream of matter which must be in a very finely divided state.

The kathode rays can be deflected by a magnet. This is obvious at once: on bringing a magnet near the stream it is deflected, just as it should be according to known laws if it were a current flowing in a perfectly flexible conductor.

A screen with a narrow horizontal slit is placed in front of the kathode ; the rays passing through the slit form a narrow horizontal beam and produce a patch of light on the wall of the tube opposite to the kathode.   As the magnet approaches the patch of light moves.   If the kathode is to the right as shewn in Fig. 269, and the lines of magnetic force pass through the paper from above downwards, the patch of light is raised, if the direction of the lines of magnetic force be reversed it is lowered.

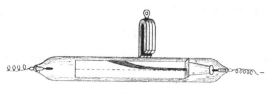

Fig. 269.

The fact that the kathode rays were carriers of negative electricity was established by J. J. Thomson.

He placed inside a Crookes' tube two metallic tubes, one within the other ; the outer one was connected to earth, the inner one was insulated and connected to an electrometer. Two narrow slits were made in the walls of the tubes opposite to each other, and by means of a magnet the kathode stream which when undeflected passed outside both tubes could be directed into the inner tube.   Here being inside a hollow closed conductor the stream gave up its electricity to the conductor, and the amount of charge acquired by the tube was measured by the electrometer.   The tube always acquired a negative charge ; the outer earthed tube served to keep off stray electrification.

If we suppose that this negative electrification is carried by particles and call n the number of the particles which enter the tube in a given time, e the charge, and m the mass of each, then an experiment such as the above enables us to find the value of ne.

262. Ionic Charge in Electrolysis.   When treating of electrolysis we were led to consider definite quantities of matter each carrying a definite charge ; to these we gave the name of ions, and we saw that the ionic charge, the quantity of electricity carried by any ion, was a constant. Moreover we have learnt that the mass of hydrogen liberated by the passage of an electromagnetic unit quantity of electricity is 1.03×10^{-4} gramme. Calling this 10^{-4} gramme, since the ratio of the mass liberated to the electricity set free is a constant for the same substance, we have if e be the charge and m the mass of a single hydrogen ion m/e equal to 10^{-4}.

Hence for hydrogen ions

$$\frac{e}{m} = 10^4.$$

If we know the mass of an ion of hydrogen we can calculate from this its charge.

263. Charge carried in Kathode Rays.

Again, if we assume e to be the charge and m the mass of the carriers of negative electricity in the kathode rays, a relation between these quantities and u the velocity with which they move can be obtained by observing the curved path of the stream in the magnetic field. When the field is at right angles to the kathode stream that path is seen to be a circle, and its radius r can be measured.

Let u be the velocity of any particle, then since m is its mass, the force which must be applied to make it describe a circle of radius r is mu^2/r. But this force arises from the magnetic field and is equal to its strength multiplied by the strength of the current. The current strength, considering a stream 1 square centimetre in section, is eu, and if H be the magnetic intensity the force tending to make the particles describe the circular paths is Heu.

Hence
$$\frac{mu^2}{r} = Heu$$

or
$$\frac{m}{e}u = Hr.$$

Now we do not know the value of u but experiment proved

Hr to have the value 3×10^6 approximately. If we assume u to be the velocity of light or 3×10^{10} cm. per second, we obtain

$$3 \times 10^{10} \frac{m}{e} = 3 \times 10^6$$

or
$$\frac{m}{e} = 10^{-4},$$

the same value as previously.

Thus the facts were consistent with the assumption that in the kathode rays, particles having the mass of the ions known to us in electrolysis were shot from the negative electrode with the velocity of light.

On the other hand many of the phenomena of the discharge in a Crookes' tube did not depend on the nature of the gas, as they should have done if the moving particles were ions, and it was difficult to imagine particles as massive as the ions moving with the velocity of light.

Hence J. J. Thomson was led to enquire further.

We have seen how he measured the total charge given up per second by the particles. Let this be Q and let the number of particles per cubic centimetre of the stream, assumed to be of unit cross section, be N. Then the number entering the inner tube per second is Nu, and their charge is Nue.

Hence we have

$$Neu = Q, \text{ and } Q \text{ is known.}$$

Again, these particles give up their energy to the inner tube; by treating the tube as a calorimeter of known heat capacity and measuring the rise of temperature per second due to the bombardment, we can find W, the energy given up by the particles. But this energy is $\frac{1}{2}Nmu^3$, for the mass of the entering particles is Nmu and the velocity of each is u.

Thus we have the three equations,

$$Neu = Q$$
$$\tfrac{1}{2}Nmu^3 = W$$
$$\frac{mu}{e} = Hr.$$

From the first two equations we obtain

$$\frac{mu^2}{e} = \frac{2W}{Q}.$$

Combining this with the equation $\frac{m}{e}u = Hr$ we find

$$u = \frac{2W}{H \cdot Q \cdot r}.$$

On substituting the experimental values on the right
hand it appears that u was about equal to 10^9 while the
value for m/e came to be about 10^{-7} instead of 10^{-4} as
obtained from electrolysis.

Thus we are bound to assume either that m is much
less, or e much greater than the corresponding electrolytic
quantities—of course both facts may be true. If however we
assume e to be the same, then m is about 1000th part of the
mass of an ion of hydrogen.

264. Number of Particles in the Kathode Rays.
The value thus obtained for the ratio m/e was substantiated
by the result of a number of other experiments. Before,
however, it is possible to determine which of the above
hypotheses as to the values of m and e separately is the true
one, it is necessary to determine N the number of particles
per unit volume.

This also has been done by Thomson. Aitken had shewn
that a cloud is ordinarily formed by the condensation of water
vapour about small nuclei of dust or other foreign material in
the atmosphere, and C. T. R. Wilson has proved that the
particles which serve as the carriers of negative electricity in
the kathode rays would serve also as such nuclei.

If now a cloud be formed, by allowing the moist air in a
closed vessel to expand suddenly, and then permitted to settle,
the nuclei are cleared away by the cloud, and so long as no
fresh nuclei are formed, no further clouds will be produced
in the space.

Thus the number of nuclei in the space will be the same
as the number of drops in the cloud.

The total volume of the drops formed can be calculated

from a knowledge of the expansion producing the cloud; hence if we can calculate in any way the size of the drops we can obtain their number by dividing the total volume by the volume of each drop.

The drops settle down at a uniform rate because the resistance due to the air just balances the weight of the drop; and a relation can be found between the velocity of a drop which has attained this terminal speed and its radius. From this relation it is possible to calculate the radius of a drop, and hence the number of drops in each cubic centimetre.

The result of J. J. Thomson's experiments was to shew that the number of drops per c.c. under the conditions of his experiment was 4×10^4, and that under the same circumstances the value of e, the charge carried by each particle, is $6 \cdot 3 \times 10^{-10}$ electrostatic units.

265. Charge of an Ion in Electrolysis. Now we have seen that the charge carried by one gramme of hydrogen in electrolysis is about 10^4 electromagnetic units of electricity; more accurately it is 9650 units; the number of hydrogen ions in a gramme can be determined by means of experiments on the viscosity of hydrogen, and is found to be $4 \cdot 4 \times 10^{23}$, and the number of electrostatic units in one electromagnetic unit is 3×10^{10}. Thus the charge on a hydrogen ion in electrostatic units is $9650 \times 3 \times 10^{10}/44 \times 10^{22}$ or $6 \cdot 5 \times 10^{-10}$ electrostatic units, the same practically as that carried by the particles in the kathode discharge.

266. Corpuscles. Again we have seen that for hydrogen ions the ratio m/e is about 10^{-4}, while for the kathode particles it is 10^{-7}; more exactly the value is $1 \cdot 4 \times 10^{-7}$. Hence the mass of the kathode ray particle is about $1 \cdot 4 \times 10^{-3}$ (or 1/700th) of that of an hydrogen ion.

J. J. Thomson calls these particles corpuscles; other writers have given them the name of electrons, and the most modern theory of electricity is based on the supposition that practically all the phenomena of electricity are due to the presence and motion of electrons.

267. Electrons. Electrons can be freed by other means than the kathode rays.

Thus Lenard shewed that if the kathode rays be allowed to fall on a window of thin aluminium sealed into the walls of a vacuum tube, rays having much the same properties penetrate into the air outside the tube; we may suppose either that some of the electrons penetrate through the aluminium or that the impact of the electrons inside the tube is sufficient to knock off electrons from the outer side into the air.

268. Rontgen Rays. Under the same circumstances as these, Rontgen had shewed that another system of rays are produced in the air.

The Rontgen or X rays pass through glass and many other materials very easily. When they fall on certain fluorescent substances they excite these strongly, they can penetrate readily substances like paper, wood, and light materials which are opaque to light. The denser metals are more opaque; their use in surgery, owing to the fact that they penetrate the muscle and lighter tissues of the body while the bones are opaque to them, is well known. They are best produced by the use of a tube of soda glass of the shape shewn in Fig. 270.

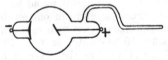

Fig. 270.

The kathode is of aluminium and is concave. The kathode rays are shot on to the platinum anode inclined at 45° to the direction in which they are travelling and from this the Rontgen rays appear.

According to the theory of Stokes, developed by Thomson, each electron as it reaches the anode sets up a sudden impulse in the ether which travels outwards as a single wave from the anode; these impulses follow each other rapidly but not with the rapidity or regularity requisite to produce a continuous stream of light waves; the properties of such a discontinuous series of impulses will resemble, as Thomson proved, those of the Rontgen rays.

269. Production of Electrons. Electrons are produced by the action of Rontgen rays; thus air through which Rontgen rays are allowed to pass loses its insulating power and becomes conducting, owing to the production of free electrons, each carrying its negative charge. They are also produced by the action of ultra-violet light on negatively charged bodies. A body when charged negatively, loses its charge rapidly when illuminated by ultra-violet light. Some substances—as salts of Uranium, Radium and certain salts of Thorium—give off radiation continuously and a part of the radiation which they emit has the properties of electrons. These radiations have been studied by M. H. Becquerel, their discoverer, M. and Madame Curie, Professors Rutherford, Townsend and others.

It is to be noticed that in all cases the electrons or corpuscles are associated with a negative charge of electricity; streams of positively charged matter may exist, but the masses concerned are comparable with those of the atoms, instead of being nearly one thousand times smaller than hydrogen atoms.

270. Electron Theory of Electricity. According to the electron theory a neutral atom consists of an electron or series of electrons each carrying its negative charge together with a positively charged nucleus, the total positive charge being equal to the sum of the negative charges on the electrons.

It is possible in various ways to attach one or more electrons to such an atom; it then becomes negatively charged; it is also by hypothesis possible to detach one or more electrons; the remainder—the coelectron as Prof. Fleming[1] has called it—remains positively electrified.

A univalent atom, like hydrogen, is one which can receive or give up one electron and no more. A divalent atom can receive or give up two electrons and so on.

---

[1] *Popular Science Monthly*, May, 1902. Prof. Fleming's article may be referred to by those who wish to learn more of the Theory, see also a lecture by Sir O. J. Lodge, delivered before the Institution of Electrical Engineers, November 27, 1902.

A current of electricity is a stream of electrons, a body through which the electrons pass freely is a conductor; within a non-conductor they cannot move about readily. A gas may be non-conducting, because of the absence of electrons, if they are introduced it gains conductivity. All the phenomena of electric discharge and current are convection phenomena.

When electromotive force is applied to a conductor, the electrons are urged through the conductor; if it be a gas at low pressure they stream from the kathode as the kathode rays.

In an electrolyte in solution, some of the free ions are positive; they are coelectrons and the electrons which have left them have joined on to other ions, making them negative; there is probably a continual interchange going on, but on the average the above statement represents the position.

The negative ions are driven by the E.M.F. to the anode, the positive ions travel to the kathode.

In a solid conductor the same kind of separation and combination of ions and electrons is taking place but the ions are not free to move; the current is conveyed by the electrons moving on from ion to ion through the solid; the solid is porous to them but not to the ions.

271. Electrons and Galvanic Action. There is in general a tendency for electrons to be set free at the common surface of any two bodies. Thus with zinc in oxygen an electric double layer is formed, the electrons in the layer of oxygen round the zinc are turned towards the zinc and are opposed by coelectrons in the outer layer of zinc; the same is true with copper, but the attraction of the zinc for the oxygen electrons is greater than that of the copper; we obtain the theory of the voltaic cell already developed.

In the case of two insulators, such as glass and silk, the conditions are the same, over each in air or oxygen there is a layer of positive electricity faced by the electrons of the oxygen.

When the glass and silk are rubbed together the action mixes up these two double layers, some electrons are dragged off the glass and retained by the silk, the one becomes positive, the other negative.

272. Electrons and Magnetism. An electric current is as we have said a convection stream of electrons. J. J. Thomson has proved mathematically that a single electron in motion produces magnetic force, the lines of force are at each moment circles whose centres lie in the direction of motion of the particle at that moment and whose planes are perpendicular to that direction.

From this follow the laws of electromagnetic action, while permanent magnets are whirls of electrons.

273. Electron Theory of Matter. We at present have but little idea as to what an electron is; we may look upon its charge as a natural unit of electricity and account for most of the phenomena of electricity with marked success by considering the results which follow from the motions of such a charged corpuscle. Some have gone so far as to express the belief that an electron is the centre of a particular kind of motion in the ether, and that matter is made up of assemblages of electrons, but the discussion of such questions would take us very wide of our limits which, indeed, have been considerably exceeded already.

ANSWERS TO EXAMPLES.

ELECTROSTATICS.

CHAPTER VI. Page 101.

15. 3,000 electrostatic units.   **16.** 90 electrostatic units.

17. (a) Force of attraction equal to 2 dynes.

    (b) Force of repulsion equal to ·25 dyne.

18. (1) The sphere c should be placed in the line of the spheres a and b; on the opposite side from a and 141·4 cms. from b.

    (2) In the line of the spheres; same side as a and 141·4 cms. from b.

19. (a) Zero position lies between the spheres at a distance of $\frac{2}{3}$ metre from more highly charged sphere.

    (b) Zero position in line of spheres; 200 cms. from more highly charged sphere, and 100 cms. from the lesser charged one.

20. $24\sqrt{g}$ electrostatic units. g = gravitational force in dynes.

23. (1) The same Q units.   (2) Potential $= Q\left(\dfrac{1}{c_1} + \dfrac{1}{c_2} + \dfrac{1}{c_3}\right)$.

25. Charges: Large sphere 6 times the charge of smaller one.

    Potentials: ,,   ,,   same potential as   ,,   ,,

    Densities: ,,   ,, $\frac{1}{6}$ the density of   ,,   ,,

    Energies: ,,   .,   6 times the energy of   ,,   ,,

26. 125 ergs.   **27.** 200 electrostatic units.

29. Loss in energy of $c = \frac{1}{2}\,cv^2\left[1 - \dfrac{c^2}{(c+c')^2}\right]$.

    Gain in energy of $c' = \dfrac{\frac{1}{2}\,c'c^2v^2}{(c+c')^2}$.

    Total loss due to redistribution of charge $= \frac{1}{2}cv^2\left(1 - \dfrac{c}{c+c'}\right)$.

30. $1 : \frac{1}{4}$.　　　　　**31.** (1) Same $= \theta$.　(2) Half $= \dfrac{\theta}{2}$.

32. Total charge　$= 36$ electrostatic units.
　　,,　capacity $= 9$　　　,,　　　　,,
　　Final potential $= 4$　　,,　　　　,,

35. 202·5 electrostatic units.

36. (a) 1279·95 units of capacity.
　　(b) Capacity increased k times, where $k =$ specific inductive capacity of turpentine.

37. (1) Difference of potential $= 2\sqrt{\pi}$ electrostatic units.
　　(2) Charge on either plate $= \dfrac{500}{\sqrt{\pi}}$　　　,,　　　　　,,

38. (1) 400 electrostatic units.　(2) 800 electrostatic units.

39. (1) 837 electrostatic units.　(2) 418·5 v^2 ergs.
　　　　(The base of the jar is assumed to be coated.)

40. 1×10^{-6} c.g.s. units of heat.

41. Difference of potential $= \sqrt{5\pi} = 3·96$.

MAGNETISM.

CHAPTER XII.　Page 180.

2. 6 dynes.　　　　　　**3.** Strength of pole $= 18$ c.g.s. units.

4. Moment of couple $= 600$.　**5.** Moment of couple $= 72$.

6. $A : B : C$ as $1 : 2 : 2$.　**7.** ·0008 dynes.

8. 14·14 dynes.　　**10.** ·0123 dynes.　　**11.** $1 : 1·21$.

12. Forces of $\dfrac{·18\,M}{l}$ dynes each, applied at the ends of the magnet;
　　($l =$ length of magnet); the couple produced by the forces to act in opposition to the couple produced by the Earth's magnetic force.

13. 120° more.　**14.** 11,250 c.g.s. units.　**15.** $1 : 1·176$.

16. 18·37 oscillations per minute.

17. $\sqrt{99·5} =$ approx. 10 oscillations per minute. (The magnet is assumed to be very long.)

18. 53·48 c.g.s. units.　　　　**19.** Tangent of angle $= ·98$.

VOLTAIC ELECTRICITY.

CHAPTER XIX. PAGE 316.

1. Oxygen $= 8\cdot24 \times 10^{-4}$.
   Iron (Ferrous) $= 28\cdot84 \times 10^{-4}$.
   Iron (Ferric) $= 19\cdot23 \times 10^{-4}$.

2. $1\cdot03 \times 10^{-4}$. 3. $1\cdot1412$ gms. 4. $3\frac{1}{8}$ gm.

5. Copper $= \cdot7875$ gm. Hydrogen $= \cdot025$ gm. Oxygen $= \cdot20$ gm.

6. (1) $\cdot03$ c.g.s. unit of current.
   (2) $8\cdot3$ volts $= 8\cdot3 \times 10^{-8}$ electromagnetic units of potential.

7. Resistance of wire $= 2$ ohms.
   ,, ,, battery $= 1\cdot2$ ohms.

8. $\cdot4$ ohms. 9. (a) $\frac{5}{12}$ ampere. (b) $12\cdot5$ volts.

10. (a) $1\cdot0$ volts. (b) $1\cdot0$ ohms. 11. (a) $1\cdot5$ volts. (b) $\cdot5$ ohms.

12. Note that Leclanché cell has greater internal resistance.

13. $\cdot75$ amps.

14. 3 rows in parallel; each row containing 4 cells in series, or
    4 ,, ,, ,, ,, 3 ,, .

16. 2 rows in parallel, each row containing 5 cells in series.

17. 2 cells in parallel, and the third in series with them.

18. $\frac{1}{12}$th amp.

19. (1) $\cdot875$ amp. (2) $52\cdot5$ coulombs. (3) $10\cdot94$ volts.

20. 2 rows in parallel, each row containing 24 cells.

21. (a) Current in wire of 1 ohm resistance $= \cdot6$ amp.
    (b) ,, ,, ,, $1\cdot5$ ohms ,, $= \cdot4$,,
    (c) ,, in cell $= 1\cdot0$ amp.

22. (1) (a) Current through cell $= \frac{1}{3}$ amp.
        (b) ,, ,, galvanometer $= \frac{1}{3}$ amp.
    (2) (a) Current through cell $= \frac{9.8}{?}$ amp.
        (b) ,, ,, galvanometer $= \frac{4.9}{?}$ amp.

23. $1:2$. 24. Resistance of $CB = 70\frac{2}{3}$ ohms. 25. $\frac{1}{100}$ volt.

26. $\cdot0501$ gms of Hydrogen. 27. $8\frac{1}{3}$ ohms. 28. $1\frac{6}{7}$ ohms.

29. 12 ohms; $7\frac{1}{3}$ ohms; $5\frac{1}{2}$ ohms; $4\frac{4}{5}$ ohms; $1\frac{1}{11}$ ohms.

30. 100,000 ohms. 34. $12\frac{2}{9}$ ohms.

35. $1\cdot3 \times 10^{-9}$ amps. for 1 div. deflexion. 36. $11\frac{1}{6}$ ohms.

37. Specific Resistance $= 2 \times 10^4$ in absolute measure.

39. $1:4$. 40. $1:\cdot747$. 41. $17\cdot32$ ohms. 42. $1:3$.

43. $\cdot33$ amp. 44. $34\cdot9$ cms. 45. $\frac{10}{\pi}$ turns. 46. 286 turns.

48. $\cdot169$ grams of copper. 49. $\cdot0182$. 51. $18\cdot86$ ohms.

53. 4670 ohms or 46·7 ohms; the value depending on the ratio coils to
      which the adjustable resistance was adjacent. If adjacent to
      the 10 ohms, then resistance measured was 4670 ohms; if to
      the 100 ohms, then resistance was 46·7 ohms.

54. 2·5 volts.   55. 40° C.   56. 2·1 ohms.

57. 4 rows in parallel; each row containing 2 cells in series.

58. Current in 1 ohm wire = $\frac{16}{21}$ amps.

      ,,   ,,   3   ,,   ,,   = $\frac{6}{21}$ amps.

      Heat in   1   ,,   ,,   = ·64 units per sec.

      ,,   ,,   3   ,,   ,,   = ·21   ,,   ,,

59. 1 : 2·49.   60. 13·416 ohms.

61. Current in thicker wire must be 2·828 times greater than in thinner
      wire.

63   ·025 Units of Heat.

65. (a) 5 Horse-power approx. (b) ·357° C. per minute.

66. 1 : ·0025.   67. (1) 6·5 Kilowatt-Hours. (2) ·042 Kilowatt-Hours.

68. 16 sq. cms.

ELECTROMAGNETISM AND INDUCTION OF CURRENTS.

CHAPTER XXIII. Page 387

1. ·32 c.g.s. unit.

ELECTROMAGNETISM AND ELECTRO-MAGNETIC INDUCTION.

CHAPTER XXIV. Page 404.

1. No effect.

2. 180° less $(\tan^{-1} ·744 + \tan^{-1} 1·524) = 87°$ approx. Note that on
     reversing the current, the needle swings through 180° owing to
     the field produced by the current being greater than H.

3. $6\frac{2}{3}$ dynes.   4. ·0157 dynes.

5. (1) $\frac{·8}{d}$ dynes. (d = perpendicular distance in cms. between the wires.)

     (2) The force is one of attraction, and acts at right angles to both
          wires, and in the plane containing them.

7. 525 c.g.s. units. 8. (1) 7238. (2) 7238μ. 9. $4·7 \times 10^{-3}$ volts.

12. 45·65 c.g.s. units. (Assumes rate of change of potential to be
      uniform.)

INDEX.

The references are to the pages.

Electrification, 4, 5; by induction, 7

Electro-chemical equivalents, 5; determination of, 250

Electrodynamometer, the, 367

Electrolysis, observations on, 193; Faraday's laws of, 194; transformations in, 304; Ionic charge in, 421, 425

Electromagnetic Action, 328, 337; between conductors, 336

Electromagnetic forces, 331

Electromagnetic machines, 392

Electromagnets, 341

Electrometer, the Attracted Disc, 90; the Quadrant, 94

Electrometers and Electroscopes, 87

Electromotive Force, in a simple circuit, 210, 211; of a Daniell's cell, 303; standards of, 204

Electron Theory, 427; of Matter, 429

Electrons, 425; production of, 427; and galvanic action, 428; and magnetism, 429

Electrophorus, the, 74

Electroscopes, 9–15; condensing, 68; and electrometers, 87

Electrostatic Actions, explanation of, 23

Electrostatic Measurement of Potential, 93

Electrostatic and Multicellular Voltmeters, 97

Energy, of a Charged Conductor, 55; of a Charged Condenser, 62

Energy, Electrical, 291; needed to magnetise iron, 356

Energy changes in a cell, 302

Equilibrium condition, 337

Equipotential Surfaces, 31, 53

Equivalent, electro-chemical, 195, 250

Experiments with Leyden Jars, 68

Experiments with the Magnetometer, 157

Experiments on electric currents, 245

Experiments on batteries, 255

Explanation of Electrostatic Actions, 23

Faraday's experiments on induction, 373

Faraday's laws of electrolysis, 194

Field of Force, 39

Field, Magnetic, 111; tensions and pressures in, 118

Force, the Field of, 39

Force, Law of, 44

Force, Lines of, 32; and equipotential surfaces, 34; forms of, 35; magnetic, 112; tracing, 114; electromotive in a circuit, 210, 211

Frictional Machines, 70

Galvanic action, 428

Galvanometer, tangent, 227; construction of, 230

Galvanometer, sine, 229

Galvanometer, resistance of, 274

Galvanometer, ballistic, 279

Galvanometer Constant, 228

Galvanometers, 223, 226; sensitive, 232; moving coil, 361

Gases, electric discharge through, 418

Gramme ring, the, 397

Grove's Cell, 202

Holtz Machine, the, 82

Hysteresis, 354

Illustrations, Mechanical, 40

Inductance, unit of, 379

Induction, electrification by, 7; magnetic, 111, 342; curve of, 352; Faraday's experiments on, 373; coefficient of mutual, 377; observations on, 379

Induction Coil, the, 389

Inductive Capacity, 45, 62

Inertia, electric, 411

Influence, of Temperature, 130; of external fields, 371

Influence Machines, 77

Insulating Medium, importance of, 30

Printed in the United States
By Bookmasters